W9-CRB-754

OXFORD STUDIES IN PHYSICS

GENERAL EDITORS

B. BLEANEY, D. W. SCIAMA, D. H. WILKINSON

THE THEORY OF
POLARIZATION
PHENOMENA

BY

B. A. ROBSON

CLARENDON PRESS · OXFORD
1974

PHYSICS

Oxford University Press, Ely House, London W.1

GLASGOW   NEW YORK   TORONTO   MELBOURNE   WELLINGTON
CAPE TOWN   IBADAN   NAIROBI   DAR ES SALAAM   LUSAKA   ADDIS ABABA
DELHI   BOMBAY   CALCUTTA   MADRAS   KARACHI   LAHORE   DACCA
KUALA LUMPUR   SINGAPORE   HONG KONG   TOKYO

ISBN 0 19 851453 0

© OXFORD UNIVERSITY PRESS 1974

All rights reserved. No part of this publication may be reproduced, stored in a retrieval system, or transmitted, in any form or by any means, electronic, mechanical, photocopying, recording or otherwise, without the prior permission of Oxford University Press

PRINTED IN GREAT BRITAIN
BY J. W. ARROWSMITH LTD, BRISTOL BS3 2NT

Q C
794
.6
S3
R61
PHYS

PREFACE

THE purpose of this book is to provide a detailed but simple development of the general formalism required to describe the polarization of particles of arbitrary spin and their decay into or interaction with other particles. The book evolved out of two series of lectures given at the Australian National University, Canberra, mainly to post-graduate students and research workers in the field of nuclear physics. There appeared to be a definite need for a book devoted entirely to the theory of polarization phenomena. While many books of a more general nature contain some discussion of the polarization of particles, in most cases the treatment is too superficial and inadequate. On the other hand the specialist papers and review articles tend to lack sufficient introduction to the subject for the average research worker to grasp the underlying physics of the often quite complicated theoretical formalism.

In this book I have tried to show the unity and logical development of the subject from the initial discovery of the polarization of light. In particular I have given considerable emphasis to the method developed by Stokes, Soleillet, Perrin, and Mueller in classical optics now known as the Mueller calculus. This approach, which offers considerable simplicity and clarity both for designing polarization experiments and for understanding the resultant polarization measurements, has been almost completely ignored outside of optics. I hope that the present book will rectify this situation.

The book is intended to be read from the beginning to the end. It has been written at the level of a graduate physics course and assumes a basic knowledge of quantum theory, scattering theory, and matrix algebra. It should be useful to research workers in all branches of physics (optics, atomic, nuclear, particle, etc.) who study polarization effects.

The book is essentially pure theory with no experimental results being discussed. Thus no attempt has been made to consider all polarization phenomena; only a few simple examples (mostly from my own field of nuclear physics) are presented in order to illustrate the theoretical formalism.

As far as possible I have adopted the Madison convention for specifying polarization quantities. Unfortunately, this convention is not consistent with the logical definition of the analysing powers for the scattering of a polarized incident beam in the spherical tensor representation. For this reason I have denoted the Mueller matrix by Z rather than T. The Madison convention does not include quantities such as polarization transfer and spin-correlation coefficients. I hope that the notation and definition of these quantities adopted in this book will find ready acceptance by research workers.

0859

I have taken particular care to differentiate between quantities which specify the polarization of a beam of particles and quantities which specify the interaction of such particles with a target. Thus I have introduced the quantity 'vector scattering parameter' for the elastic scattering of spin-$\frac{1}{2}$ particles by a spinless target. Although this quantity is sometimes numerically equal to the vector polarization of the scattered beam, it is confusing to identify the two quantities too closely.

I wish to express my gratitude to Dr. N. Berovic who read almost all of the manuscript and gave me very valuable criticism and advice. I am also indebted to many colleagues for discussions and comments on the manuscript and Mrs. I. Kinchin for preparing most of the final manuscript in such a cheerful and competent manner.

I am grateful to Profs. W. E. Burcham and F. Beck for hospitality at the University of Birmingham and the Institut für Kernphysik, Darmstadt, respectively, where considerable portions of the book were written. I also thank Prof. K. J. LeCouteur for encouragement to commence such a project.

B.A.R.

The Australian National University, Canberra
July 1974

CONTENTS

INTRODUCTION

THE observation by Bartholinus in 1669 of the double refraction of light by calcite led to the discovery of the *polarization* of light by Huygens *c.* 1690. Huygens found that light which had passed through a piece of calcite behaved differently from ordinary light. However, although he was able to describe the phenomenon of double refraction in terms of his wave construction, he was unable to account for the polarization of light. This was not achieved until *c.* 1817 when Young suggested that light waves are transverse rather than longitudinal vibrations (Fresnel claimed to have mentioned this possibility to Ampère in 1816). In 1824 Fresnel showed that light waves are exclusively transverse and the resultant transverse vector theory constituted the first theory of polarized light. This was eventually superseded by the more general electromagnetic theory of Maxwell in 1864.

The term *polarization* was first used by Malus in 1810 when describing the production of polarized light by reflection and was derived from the word 'polarity' employed much earlier to describe the two-sidedness or two-fold nature of magnetic poles. Malus, while keeping an open mind on the wave-versus-corpuscular theories of light, employed the latter model and considered that the polarization of light was connected with the polarity of the corpuscles. In this sense the word polarization is a misnomer but the term has such a long history that there can be no question of a replacement for it. Unfortunately, polarization has also been used to describe other effects, e.g. in Maxwell's displacement vector $\mathbf{D} = \mathbf{E} + 4\pi\mathbf{P}$. Here $\mathbf{E}$ is the electric vector and the polarization vector $\mathbf{P}$ is a measure of the mean polarizability of a dielectric medium. We do not consider such phenomena in this treatise.

The next polarization phenomena to be observed and described were essentially the Zeeman effect and the doublet spectral lines of the alkali elements. In 1896 Zeeman discovered that certain spectral lines are split into a number of components on the application of an external magnetic field. The classical theory of Lorentz indicated the splitting of lines into three components (the normal Zeeman effect) and indeed in some cases is able to account for the measurements. However, in many instances there occur more than three components—the so-called anomalous Zeeman effect. For the explanation of this latter phenomenon and the alkali doublet structure, it was necessary to assume that the electron possesses an intrinsic angular momentum called spin, which by analogy with the quantization of orbital angular momentum gives rise to just two basic states, i.e. electrons are spin-$\frac{1}{2}$ particles. The spin of the electron gives rise to a small splitting of the majority of alkali atomic energy levels and the corresponding occurrence of doublet spectral lines. Associated with the spin angular momentum is a magnetic moment which accounts for the anomalous Zeeman effect.

Since 1925, when the concepts of electron spin and magnetic moment were published, each particle or system of particles (e.g. deuteron) is considered to possess a spin s which has a unique value from the set $0, \frac{1}{2}, 1, \frac{3}{2}, \ldots$. Thus pions and alpha particles have $s = 0$, electrons and protons have $s = \frac{1}{2}$, photons and deuterons have $s = 1$, etc. All particles with $s > 0$ exhibit polarization phenomena, i.e. effects which arise as a direct consequence of their intrinsic spin. This book deals only with the description of such polarization phenomena.

For several reasons we discuss polarized light first (Chapter 1). Since the polarization of light was studied for over two hundred years before any other polarization phenomenon, much of the terminology of polarization theory is derived from optics. Secondly, polarized light offers simple and convenient examples with which to introduce the various formalisms. Moreover, this can be done using a classical approach which at a later stage and as a separate step may be simply re-interpreted in a quantum-theory treatment (Chapter 2). The two main approaches now employed in optics for describing the inter-action of polarized light with optical devices are the Jones and Mueller calculi, which were invented in the early 1940s. Both these matrix methods are conveniently introduced by considering the passage of polarized light through two simple optical instruments, the 'stopped' calcite crystal (or its equivalent the Nicol prism) and the quarter-wave plate, which have compara-tively trivial matrix representations. Furthermore, from their study of classical optics, many readers undoubtedly will be familiar with the properties of these devices as well as the different forms of polarized light. Finally, the photon concept played a leading role in the development of quantum theory and presents a convenient stepping stone for the transition to a general description of polarized particles.

The two calculi employed in classical optics have their counterparts in the description of the polarization of particles of arbitrary spin and their decay or interaction with other particles. Indeed rather earlier than 1940, the equivalent of the Jones calculus, namely, the use of spin wave functions and scattering (or reaction) matrices was introduced into particle physics by Pauli in 1927 and Wheeler in 1937, respectively. The extension of the Jones calculus to include partially polarized light requires the use of the density matrix proposed by von Neumann in 1927. On the other hand, the Mueller method has scarcely been employed for particles. However, it is the author's belief that this complementary approach offers considerable simplicity com-pared with the usual description based upon density and reaction matrices. Both methods and the relationship between them are discussed in detail (Chapters 3–6).

In Chapters 3 and 4 we discuss in depth the complete specification of the polarization of spin-$\frac{1}{2}$ and spin-1 particles, respectively. Emphasis is also given to spin-1 particles since they exhibit fundamental differences from the simplest case of spin-$\frac{1}{2}$ particles which are typical of higher spin particles. Our over-all approach is to proceed by analogy and induction from the simpler to the more complicated phenomena rather than to commence with a general formalism. This necessitates some repetition, but the author believes that this gradual development of the theory is more understandable. Thus only the simplest reaction, elastic scattering from spinless targets, is considered at this stage. The general forms of the elastic scattering matrices under certain invariance requirements (e.g. parity conservation) and the corresponding numbers of independent observable quantities are discussed. Throughout we employ a notation to denote reference axes which is becoming more essential as the trend now is to refer initial and final spin states to different coordinate frames. We use a set of 'standard' axes which correspond to the helicity coordinate systems of Ohlsen (1972).

In Chapter 5 we present the general non-relativistic formalism for particles of arbitrary spin interacting with targets with spin. Both non-elastic reactions and processes involving identical particles are discussed. The formalism is then extended to the emission and absorption of electromagnetic radiation (Chapter 6). The concepts of decay and absorption matrices are introduced and it is believed that the treatment of angular correlations presented here is both novel and simpler than previous descriptions.

Finally we include a short chapter on the treatment of relativistic particles. We follow the approach of Chou and Shirokov (1958) which represents (at least to the author) the only correct explicit treatment of the spin precession for an arbitrary proper Lorentz transformation. We indicate how this relativistic rotation of the spin may be easily incorporated into the non-relativistic formalism. We also show the relation of our formalism to the helicity representation of Jacob and Wick (1959). For various reasons we have not adopted the helicity formalism from the outset but in the case of particles with mass this is shown to be no disadvantage for the standard axes chosen. The case of massless particles is also considered.

1

POLARIZED LIGHT

1.1. Calcite crystal experiment

CALCITE is a rhombohedral crystalline form of calcium carbonate in which
the double refraction of light is very strikingly exhibited. In a perfectly formed
crystal (Fig. 1.1) the rhombohedron is bounded by six similar parallelograms
the obtuse angles of which are about 101° 55'. The solid angles at the corners
A and G are contained by three obtuse angles while the remainder are
bounded by one obtuse and two acute angles. The plane ACGE and the line
AG are called the principal plane and the principal axis respectively. A
calcite crystal is said to be uniaxial since there is one special direction called
the optic axis in which only single refraction occurs. For calcite the optic
axis coincides with the direction of the principal symmetry axis AG.

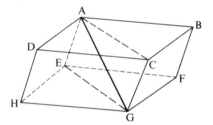

FIG. 1.1. Perfectly formed calcite crystal with principal plane ACGE and principal (optic)
axis AG.

When a beam of ordinary light is passed through a slab of calcite crystal
each ray is generally divided into two, an ordinary (O) ray which obeys the
usual law of refraction and a so-called extraordinary (E) ray which does not.
This phenomenon is undoubtedly related to the crystalline structure of
calcite since substances such as glass which have an irregular structure do
not exhibit such an effect. For convenience let us assume that the incident
direction is not too close to the optic axis and is both normal to the crystal
face and in a principal plane, i.e. in a direction parallel to the principal plane
(Fig. 1.2). The O-ray passes straight through while the E-ray is deflected
along a principal plane and emerges parallel to the incident ray. Thus a
rotation of slab 1 about the incident direction causes an equal rotation of the
E-ray about the same direction. When these two rays strike the face of slab 2
and provided both crystals have identical orientation of their optic axes, the
O-ray continues undeflected while the E-ray is further displaced. This shows
that light which has traversed a calcite crystal is different from ordinary light.

The light is said to be *polarized* and *unpolarized* respectively. Furthermore, the O- and E-rays are different so that light can exist in at least two different polarized states. However, it is necessary to be careful here; one should not visualize light as an incoherent mixture of two kinds X and Y, which are separated by the calcite slab so that the O-ray consists of type X and the E-ray of type Y. If this were so, it would not be possible to obtain a further separation in the second crystal as does in fact occur if the orientations of the optic axes are different (Fig. 1.3). In this case the calcite slab transforms the set of states $[X, Y]$ into other polarization states $[1, 2, 3, 4]$ depending upon its orientation α.

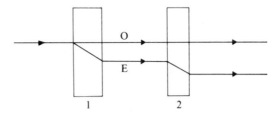

FIG. 1.2. Double refraction of ordinary (unpolarized) light by two successive calcite crystals having the same orientation of optic axes. Both the ordinary (O) and extraordinary (E) rays are polarized. All rays are in the plane of the page.

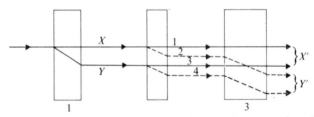

FIG. 1.3. Double refraction of ordinary (unpolarized) light by three successive calcite crystals with crystals 2 and 3 having the same orientation of optic axes. The X and Y rays each separate into two components in crystal 2. Rays 1 and 3 (and likewise rays 2 and 4) behave similarly on passing crystal 3. Rays 2 and 4 represented by broken lines are not in the plane of the page.

If the four rays are passed through a third calcite crystal which has its optic axis in the same direction as the second crystal, it is found that rays 1 and 3 behave similarly and also that rays 2 and 4 are alike. Thus there are only two types of polarization states again. Indeed, the interaction of light with all types of optical devices can be satisfactorily described in terms of just two states of polarization. It is therefore possible to write the effect of a calcite crystal upon polarized light as

$$[X', Y'] = T(\alpha)[X, Y] \qquad (1.1)$$

which means that the polarization states X, Y are transformed by the operator $T(\alpha)$, which describes in some manner the orientation and action of the crystal, into the polarization states X', Y'. This form of relationship immediately suggests that light cannot be a scalar quantity; it must have components. The phenomenon of polarization shows that a satisfactory theory of light has to be vectorial in character, i.e. be concerned with directions. It should be noted that if one starts with ordinary light two calcite crystals are required to observe the polarization effect, the first to polarize the light and the second to analyse it. One speaks of a *polarizer* and an *analyser* respectively and of course a polarizer can act as an analyser and vice versa.

1.2. Electromagnetic theory of light

The electromagnetic theory represents light in free space as transverse vibrations of electric and magnetic fields **E**, **H** which are mutually perpendicular and in phase. For the purposes of the discussion here, it is only necessary to consider one of the vectors, say **E**, since the vectors are related to each other in terms of simple constants so that **H** can be derived from **E**.

The two rays of light transmitted by a calcite crystal are both *linearly polarized*, i.e. the electric vectors representing the rays have fixed directions so that each vibration takes place in a single plane containing the direction of propagation. Moreover, these two directions are at right angles to one another. It has been found that the polarization of light can be understood in terms of the superposition of two such rays; a given polarized ray of light may be represented as the resultant of two disturbances, one with the **E**-vector in the xz-plane and the other with the **E**-vector in the yz-plane and both travelling along the z-axis. We can write

$$\mathbf{E} = E_x \mathbf{e}_x + E_y \mathbf{e}_y, \tag{1.2}$$

where

$$E_x = a\cos(kz - \omega t) = a\cos\phi, \tag{1.3}$$

and

$$E_y = b\cos(kz - \omega t + \delta) = b\cos(\phi + \delta). \tag{1.4}$$

Here $\mathbf{e}_x$, $\mathbf{e}_y$ are unit vectors along the x-, y-axes, ω is the angular frequency, and k is the wave number. The two component vibrations have the same frequency and velocity of propagation but their amplitudes differ and there is a permanent phase difference δ. In general, the tip of the **E**-vector will appear to trace out an ellipse (the polarization ellipse) when viewed along the direction in which the light propagates, and consequently a polarized ray of light is said to be *elliptically polarized*. If the **E**-vector rotates around the ellipse in a clockwise (anti-clockwise) direction when viewed by an observer who receives the beam of light, the ray is said to be right-handed (left-handed) elliptically polarized light. There are two special cases: (1) if $a = b$ and $\delta = \frac{1}{2}m\pi$ $(m = \pm 1, \pm 3, \pm 5, ...)$ the ellipse becomes a circle and the light is called right-handed and left-handed *circularly polarized* light respectively; (2) if $\delta = m\pi$ $(m = 0, \pm 1, \pm 2, ...)$ the ellipse degenerates to a straight line (the polarization line) and the light *is linearly polarized*.

1.3. Stokes parameters

To specify the polarization ellipse we require three independent quantities, e.g. the amplitudes a and b and the phase difference δ. For practical purposes it is convenient to characterize the state of polarization by certain parameters which are all of the same physical dimensions and which were introduced by Stokes (1852). For a plane monochromatic wave the *Stokes parameters* are the four quantities:

$$I = a^2 + b^2, \qquad P_1 = a^2 - b^2, \qquad P_2 = 2ab\cos\delta, \qquad P_3 = 2ab\sin\delta. \quad (1.5)$$

Only three of these quantities are independent since

$$I^2 = P_1^2 + P_2^2 + P_3^2. \quad (1.6)$$

The Stokes parameters are useful because they (1) can be determined by simple experiments, (2) allow treatment of unpolarized and partially polarized beams of light, and (3) may be re-interpreted in terms of the quantum theory of light. The parameter I is a measure of the *time-averaged intensity* of the wave, the average being taken over a period long enough to allow adequate measurement but very long indeed compared with the natural period $\omega \sim 10^{-15}\,\mathrm{s}^{-1}$ of the wave. The parameters P_1, P_2, and P_3 can be expressed in terms of quantities which describe the polarization ellipse, ψ the angle between the major axis and the x-axis, and $\chi = \tan^{-1}(\pm B/A)$, $-\frac{1}{4}\pi \leqslant \chi \leqslant \frac{1}{4}\pi$, where A and B are the lengths of the major and minor semi-axes (Fig. 1.4). The quantity $\tan\chi$ is defined to have the same sign as $\sin\delta$ and is positive

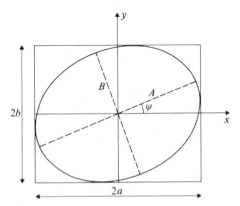

FIG. 1.4. Polarization ellipse for electric vector $\mathbf{E} = a\cos\phi\,\mathbf{e}_x + b\cos(\phi + \delta)\mathbf{e}_y$. The major and minor semi-axes have lengths A and B, respectively; ψ is the angle between the major axis and the x-axis.

(negative) for right-handed (left-handed) polarization. We have

$$P_1 = I \cos 2\chi \cos 2\psi, \tag{1.7a}$$

$$P_2 = I \cos 2\chi \sin 2\psi, \tag{1.7b}$$

$$P_3 = I \sin 2\chi. \tag{1.7c}$$

These relationships bear a close resemblance to the formulae for the three components of a vector expressed in spherical coordinates and indicate a simple geometrical representation of all the different states of polarization. This representation is known as the *Poincaré sphere* (Poincaré 1892).

The Poincaré sphere Σ is a sphere of radius I (Fig. 1.5) such that any point P on the surface having spherical angular coordinates $(\frac{1}{2}\pi - 2\chi)$ and 2ψ represents one and only one state of polarization of a plane monochromatic wave. The reverse is also true; each point on the surface Σ uniquely defines one state of polarization. In particular:

(1) all points 'north' ('south') of the equator represent states of right-handed (left-handed) polarization;
(2) the north (south) pole represents right-handed (left-handed) circular polarization;
(3) the equator represents all states of linear polarization.

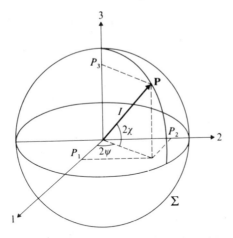

FIG. 1.5. Poincaré sphere (Σ) representation of all polarization states of a plane monochromatic wave. The polarization vector $\mathbf{P} = (P_1, P_2, P_3)$ has length I and polar angles $(\frac{1}{2}\pi - 2\chi)$, 2ψ.

1.4. Modern theories of polarized light

The description of the interaction of polarized light with several optical devices using conventional algebraic and trigonometric methods is a very difficult and cumbersome process. Two 'modern' methods which greatly simplify the description are the *Jones* and *Mueller calculi*. Both methods use matrix methods and depend upon representing: (1) the initial beam of light by a matrix $V^{(0)}$, (2) the interaction of an optical device with a beam of light by a matrix T, and (3) the light produced by the optical device T for incident light $V^{(0)}$ by a matrix $V^{(1)}$ which is the product of T and $V^{(0)}$. We write eqn (1.1) as

$$V^{(1)} = TV^{(0)}. \tag{1.8}$$

Thus the outcome of an experiment involving n successive devices, represented by $T_1, T_2, T_3, ..., T_n$ respectively, is given by simple matrix algebra:

$$V^{(n)} = T_n T_{n-1} \cdots T_3 T_2 T_1 V^{(0)}. \tag{1.9}$$

1.5. Jones calculus

In 1940–1 Jones (1941) (also Shurcliff (1962)) developed a calculus in which a plane monochromatic ray of light travelling along the z-axis is represented by a column matrix whose elements are essentially the two components E_x, E_y of the electric vector:

$$V_c = \begin{bmatrix} E_x \\ E_y \end{bmatrix}_c = \mathrm{Re} \begin{bmatrix} a\,e^{i\phi} \\ b\,e^{i(\phi+\delta)} \end{bmatrix}_c, \tag{1.10}$$

where Re stands for 'real part of' and the subscript c denotes the reference system of axes x, y, z defined by the orthogonal unit vectors $\mathbf{e}_x$, $\mathbf{e}_y$, and $\mathbf{e}_z$. Following Jones we drop the symbol Re and adopt the full complex form:

$$V_c = \begin{bmatrix} a\,e^{i\phi} \\ b\,e^{i(\phi+\delta)} \end{bmatrix}_c = e^{i\phi} \begin{bmatrix} a \\ b\,e^{i\delta} \end{bmatrix}_c, \tag{1.11}$$

it being understood that the physical displacements of the **E**-vector are given by the real parts of the two elements. The time-averaged intensity of the beam is proportional to the sum of the squares of the magnitudes of the two elements, i.e.

$$I = C(a^2 + b^2). \tag{1.12}$$

For simplicity we assume $C = 1$. If the details of the time variation are of no interest we can also omit the factor $e^{i\phi}$ and represent the state of polarization of a beam of light by the *Jones vector*

$$J_c = \begin{bmatrix} a \\ b\,e^{i\delta} \end{bmatrix}_c. \tag{1.13}$$

Thus light linearly polarized along the x-axis and a right-handed circularly polarized beam are represented by $\begin{bmatrix} a \\ 0 \end{bmatrix}_c$ and $\begin{bmatrix} a \\ ia \end{bmatrix}_c$ respectively.

The interaction operator T is assumed to be a 2×2 matrix which is appropriate for transforming polarized beams represented by Jones vectors into other polarized beams. We now construct the T operator for an *idealized* calcite crystal, i.e. one which does not absorb or reflect any of the incident light. The physical features of such a crystal orientated as in Fig. 1.2 are:
(1) the O- and E-rays are linearly polarized at right angles to one another and the directions of these polarizations rotate as the crystal is rotated about the direction of the O-ray (z-axis);
(2) the introduction of a phase difference ε between the O- and E-rays arising from the different optical paths traversed;
(3) the physical separation of the O- and E-rays if the crystal is thick enough.
It is convenient to refer the final polarization state to axes $x'y'z'$ which differ from the initial system xyz and which are defined by the linear polarizations of the emerging O- and E-rays. Choosing $\mathbf{e}_{z'} \equiv \mathbf{e}_z$, the $\mathbf{e}_{x'}$ and $\mathbf{e}_{y'}$ are related to the $\mathbf{e}_x$ and $\mathbf{e}_y$ unit vectors by a simple clockwise rotation α about the z-axis. Thus we can write

$$T_{c'c} = \begin{bmatrix} \cos\alpha & \sin\alpha \\ -\sin\alpha & \cos\alpha \end{bmatrix}_{c'c}, \tag{1.14}$$

where the subscripts c', c remind us of the change in reference axes and the angle $\alpha < \pi$ defines the rotation of axes required to bring either the x'- or y'-axis along the line of polarization of the emergent O- or E-ray, respectively. The phase difference ε is included in T as a factor $e^{i\varepsilon}$ on the second row, and the separation of the two components is symbolized by a dashed line between the two rows. Thus we have

$$T_{c'c} = \begin{bmatrix} \cos\alpha & \sin\alpha \\ \hline -e^{i\varepsilon}\sin\alpha & e^{i\varepsilon}\cos\alpha \end{bmatrix}_{c'c}. \tag{1.15}$$

For an arbitrary elliptically polarized incident ray, the resultant rays of light are given by

$$\begin{bmatrix} \cos\alpha & \sin\alpha \\ \hline -e^{i\varepsilon}\sin\alpha & e^{i\varepsilon}\cos\alpha \end{bmatrix}_{c'c} \begin{bmatrix} a \\ b\,e^{i\delta} \end{bmatrix}_c$$

$$= \begin{bmatrix} a\cos\alpha + b\sin\alpha\,e^{i\delta} \\ 0 \end{bmatrix}_{c'} + \begin{bmatrix} 0 \\ -a\sin\alpha\,e^{i\varepsilon} + b\cos\alpha\,e^{i(\varepsilon+\delta)} \end{bmatrix}_{c'}, \tag{1.16}$$

where we have mathematically separated the O- and E-rays according to our convention. We now consider two simpler special optical devices which we shall require later.

Stopped calcite crystal. This is an ideal calcite crystal orientated as in

Fig. 1.2 but with the E-component blocked out (see Fig. 1.6). The appropriate matrix is

$$T^S_{c'c} = \begin{bmatrix} \cos \alpha & \sin \alpha \\ 0 & 0 \end{bmatrix}_{c'c}. \tag{1.17}$$

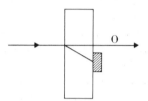

FIG. 1.6. Ideal stopped calcite crystal. The extraordinary ray is stopped.

Quarter-wave plate. If the piece of calcite is so thin that the two components are not separated and the optical path difference is a quarter of a wavelength, i.e. $\varepsilon = \frac{1}{2}\pi$, we have a device called the quarter-wave plate for which the transformation matrix is (for an ideal crystal),

$$T^Q_{c'c} = \begin{bmatrix} \cos \alpha & \sin \alpha \\ -\mathrm{i} \sin \alpha & \mathrm{i} \cos \alpha \end{bmatrix}_{c'c}. \tag{1.18}$$

As examples of the use of the Jones calculus we now consider the inter-action of these two devices with both linearly and circularly polarized light.

1.5.1. *Analysis using stopped calcite crystal*

(a) First let us consider a linearly polarized beam in which $E_y = 0$, i.e.

$$J^{(0)}_c = \begin{bmatrix} a \\ 0 \end{bmatrix}_c. \tag{1.19}$$

If we pass such a beam through a stopped calcite crystal we find by simple matrix algebra that the final state is given by

$$J^{(1)}_{c'} = T^S_{c'c} J^{(0)}_c = \begin{bmatrix} \cos \alpha & \sin \alpha \\ 0 & 0 \end{bmatrix}_{c'c} \begin{bmatrix} a \\ 0 \end{bmatrix}_c = \begin{bmatrix} a \cos \alpha \\ 0 \end{bmatrix}_{c'}. \tag{1.20}$$

Thus the resultant beam is linearly polarized along the new x'-axis ($\mathbf{e}_{x'}$) and the intensity $I^{(1)} = a^2 \cos^2 \alpha = I^{(0)} \cos^2 \alpha$ is a function of the orientation α of the crystal. This intensity variation as the crystal is rotated about the z-axis tells us that the incident light is polarized, and by analysing the effect in detail we can learn something about the nature of the initial polarization state.

(b) Secondly let us consider right-handed circularly polarized light, i.e. $a = b, \delta = \pi/2$, so that

$$J_c^{(0)} = \begin{bmatrix} a \\ ia \end{bmatrix}_c, \tag{1.21}$$

which has intensity $I^{(0)} = 2a^2$. Passing this light through a stopped calcite crystal gives

$$J_{c'}^{(1)} = \begin{bmatrix} \cos\alpha & \sin\alpha \\ 0 & 0 \end{bmatrix}_{c'c} \begin{bmatrix} a \\ ia \end{bmatrix}_c = \begin{bmatrix} a\,e^{i\alpha} \\ 0 \end{bmatrix}_{c'}. \tag{1.22}$$

Thus the resultant light is again linearly polarized but the intensity $I^{(1)} = a^2 = \frac{1}{2}I^{(0)}$ has been halved and is independent of the orientation of the crystal.

1.5.2. Analysis using quarter-wave plate

We now repeat the same experiments with a quarter-wave plate.
(a) For the linearly polarized beam we find

$$J_{c'}^{(1)} = T_{c'c}^Q J_c^{(0)} = \begin{bmatrix} \cos\alpha & \sin\alpha \\ -i\sin\alpha & i\cos\alpha \end{bmatrix}_{c'c} \begin{bmatrix} a \\ 0 \end{bmatrix}_c = \begin{bmatrix} a\cos\alpha \\ -ia\sin\alpha \end{bmatrix}_{c'}, \tag{1.23}$$

so that the emergent light is generally elliptically polarized and has intensity $I^{(1)} = I^{(0)}$.
(b) In the case of the circularly polarized beam we have

$$J_{c'}^{(1)} = \begin{bmatrix} \cos\alpha & \sin\alpha \\ -i\sin\alpha & i\cos\alpha \end{bmatrix}_{c'c} \begin{bmatrix} a \\ ia \end{bmatrix}_c = \begin{bmatrix} a\,e^{i\alpha} \\ -a\,e^{i\alpha} \end{bmatrix}_{c'}, \tag{1.24}$$

which is linearly polarized light with intensity $I^{(1)} = I^{(0)}$.

1.6. Mueller calculus

This method was developed by Mueller (Shurcliff 1962) and is based upon the Stokes parameters. As mentioned in Section 1.3, Stokes had suggested that the polarization state of a beam of light could be uniquely described by four quantities I, P_1, P_2, and P_3 according to his *principle of optical equivalence*: 'beams of light which have the same Stokes parameters are indistinguishable as regards intensity, degree of polarization and polarization form'. In 1929 Soleillet discovered that the effect of an optical device upon an incident beam of light was a *linear* transformation of the Stokes parameters and this was expressed in matrix form by Perrin in 1942. Mueller represented various optical devices by 4×4 matrices acting upon the *Stokes vector*, a column matrix whose elements are the Stokes parameters, and worked out many problems previously considered intractable.

In the Mueller calculus, a plane monochromatic beam of light is represented by a column matrix

$$
S_c = \begin{bmatrix} I \\ P_1 \\ P_2 \\ P_3 \end{bmatrix}_c = \begin{bmatrix} a^2+b^2 \\ a^2-b^2 \\ 2ab\cos\delta \\ 2ab\sin\delta \end{bmatrix}_c, \tag{1.25}
$$

where I is the time-averaged intensity of the beam, P_1, P_2, P_3 describe the form of the polarization, and the subscript c again denotes the reference axes x, y, z. The parameters P_1, P_2, P_3 measure the preference for linear polarization along the x-axis, linear polarization at $+45°$ to the x-axis and right-handed circular polarization, respectively. From eqn (1.6) we have

$$
-I \leqslant P_n \leqslant I \quad \text{for} \quad n = 1, 2, 3, \tag{1.26}
$$

and positive values of P_n indicate a preference for the above polarization form while negative values give the preference for the opposite (or orthogonal) polarization type. Thus, for $P_1 = I$,

$$
S_c = \begin{bmatrix} I \\ I \\ 0 \\ 0 \end{bmatrix}_c, \tag{1.27}
$$

which describes a beam linearly polarized along the x-axis, and for $P_1 = -I$

$$
S_c = \begin{bmatrix} I \\ -I \\ 0 \\ 0 \end{bmatrix}_c, \tag{1.28}
$$

which represents light linearly polarized along the y-axis, i.e. orthogonal to eqn (1.27). Similarly, if $P_3 = \pm I$ we have right-handed or left-handed circularly polarized light respectively.

Using the relations (1.5) it is easy to construct the appropriate Stokes vector once the corresponding Jones vector is known. Let us consider a few examples.

1. A beam of light which is linearly polarized along the x-axis is represented in the Jones calculus by $\begin{bmatrix} a \\ 0 \end{bmatrix}_c$. In the Mueller calculus this becomes

$$\begin{bmatrix} a^2 \\ a^2 \\ 0 \\ 0 \end{bmatrix}_c . \qquad (1.29)$$

2. For light linearly polarized along the y-axis $\begin{bmatrix} 0 \\ b \end{bmatrix}_c$ we have

$$\begin{bmatrix} b^2 \\ -b^2 \\ 0 \\ 0 \end{bmatrix}_c . \qquad (1.30)$$

3. Light which is linearly polarized at $+45°$ with respect to the x-axis $\begin{bmatrix} a \\ a \end{bmatrix}_c$ is described by

$$\begin{bmatrix} 2a^2 \\ 0 \\ 2a^2 \\ 0 \end{bmatrix}_c . \qquad (1.31)$$

4. For right-handed circularly polarized light $\begin{bmatrix} a \\ ia \end{bmatrix}_c$, we have

$$\begin{bmatrix} 2a^2 \\ 0 \\ 0 \\ 2a^2 \end{bmatrix}_c . \qquad (1.32)$$

As the Stokes vectors have four components, a 4×4 matrix is required to represent their transformation by an optical device. If the transformation matrix of the Jones calculus $T_{c'c}$ is known, it is possible to construct (see Section 1.8.1) the corresponding 4×4 matrix $Z_{c'c}$ in the Mueller calculus.

For the stopped calcite crystal we find

$$Z^S_{c'c} = \tfrac{1}{2}\begin{bmatrix} 1 & C_2 & S_2 & 0 \\ 1 & C_2 & S_2 & 0 \\ 0 & 0 & 0 & 0 \\ 0 & 0 & 0 & 0 \end{bmatrix}_{c'c},$$ (1.33)

where $C_2 = \cos 2\alpha$, $S_2 = \sin 2\alpha$ and α is the same angle used previously to describe the orientation of the crystal. For a quarter-wave plate we have

$$Z^Q_{c'c} = \begin{bmatrix} 1 & 0 & 0 & 0 \\ 0 & C_2 & S_2 & 0 \\ 0 & 0 & 0 & -1 \\ 0 & -S_2 & C_2 & 0 \end{bmatrix}_{c'c}.$$ (1.34)

As examples of the use of the Mueller calculus, we now consider the same experiments with these two devices which we described previously (Sections 1.5.1 and 1.5.2) in terms of the Jones calculus.

1.6.1. *Analysis using stopped calcite crystal*

(a) For light of intensity $I^{(0)} = a^2$ linearly polarized along the x-axis, the resultant beam is described by the product of $Z^S_{c'c}$ and the appropriate S_c, i.e.

$$\tfrac{1}{2}\begin{bmatrix} 1 & C_2 & S_2 & 0 \\ 1 & C_2 & S_2 & 0 \\ 0 & 0 & 0 & 0 \\ 0 & 0 & 0 & 0 \end{bmatrix}_{c'c}\begin{bmatrix} a^2 \\ a^2 \\ 0 \\ 0 \end{bmatrix}_c = \tfrac{1}{2}\begin{bmatrix} a^2(1+C_2) \\ a^2(1+C_2) \\ 0 \\ 0 \end{bmatrix}_{c'} = \begin{bmatrix} a^2\cos^2\alpha \\ a^2\cos^2\alpha \\ 0 \\ 0 \end{bmatrix}_{c'}.$$ (1.35)

The emergent beam is linearly polarized along the new x'-axis ($e_{x'}$) and has intensity $I^{(1)} = a^2\cos^2\alpha = I^{(0)}\cos^2\alpha$ as given by the Jones calculus.

(b) For right-handed circularly polarized light we have

$$\tfrac{1}{2}\begin{bmatrix} 1 & C_2 & S_2 & 0 \\ 1 & C_2 & S_2 & 0 \\ 0 & 0 & 0 & 0 \\ 0 & 0 & 0 & 0 \end{bmatrix}_{c'c}\begin{bmatrix} 2a^2 \\ 0 \\ 0 \\ 2a^2 \end{bmatrix}_c = \begin{bmatrix} a^2 \\ a^2 \\ 0 \\ 0 \end{bmatrix}_{c'}.$$ (1.36)

The resultant beam is linearly polarized along the x'-axis and has intensity $I^{(1)} = \tfrac{1}{2}I^{(0)}$.

1.6.2 *Analysis using quarter-wave plate*

(a) For an incident beam linearly polarized along the x-axis and intensity a^2 the resultant beam is given by $S_{c'}^{(1)} = Z_{c'c}^Q S_c^{(0)}$, i.e.

$$\begin{bmatrix} 1 & 0 & 0 & 0 \\ 0 & C_2 & S_2 & 0 \\ 0 & 0 & 0 & -1 \\ 0 & -S_2 & C_2 & 0 \end{bmatrix}_{c'c} \begin{bmatrix} a^2 \\ a^2 \\ 0 \\ 0 \end{bmatrix}_c = \begin{bmatrix} a^2 \\ a^2 C_2 \\ 0 \\ -a^2 S_2 \end{bmatrix}_{c'}, \qquad (1.37)$$

which is generally a mixture of linearly polarized light ($P_1 = a^2 C_2$) and circularly polarized light ($P_3 = -a^2 S_2$), i.e. is elliptically polarized and has intensity $I^{(1)} = I^{(0)} = a^2$.

(b) For right-handed circularly polarized light we obtain

$$\begin{bmatrix} 1 & 0 & 0 & 0 \\ 0 & C_2 & S_2 & 0 \\ 0 & 0 & 0 & -1 \\ 0 & -S_2 & C_2 & 0 \end{bmatrix}_{c'c} \begin{bmatrix} 2a^2 \\ 0 \\ 0 \\ 2a^2 \end{bmatrix}_c = \begin{bmatrix} 2a^2 \\ 0 \\ -2a^2 \\ 0 \end{bmatrix}_{c'}, \qquad (1.38)$$

which is light linearly polarized at $-45°$ to the x'-axis with intensity $I^{(1)} = I^{(0)} = 2a^2$.

In both the Jones and the Mueller calculus, the action of a set of optical devices is found by multiplying all the appropriate matrices together to determine the equivalent 2×2 or 4×4 matrix, e.g.

$$Z_{c^nc}^{(\text{equiv})} = Z_{c^nc^{n-1}}^{(n)} \dots Z_{c''c'}^{(2)} Z_{c'c}^{(1)}. \qquad (1.39)$$

1.7. Unpolarized and partially polarized light

Thus far we have constructed matrix representations for all types of polarized beams and also for the corresponding transformation operators of both a stopped calcite crystal and a quarter-wave plate (assuming ideal crystals). We now consider the representation of unpolarized light. Such a beam can be incorporated naturally into the Mueller formalism and we discuss this first.

Ordinary light is generally unpolarized, and when such light is passed through a stopped calcite crystal the emergent beam is linearly polarized and has an intensity $\frac{1}{2}$ of the incident beam irrespective of the orientation of the crystal. In this respect unpolarized light behaves like circularly polarized light (Section 1.6.1). However, if unpolarized light is passed through a quarter-wave plate, the resultant beam is unpolarized while a circularly polarized beam becomes linearly polarized (Section 1.6.2). The equivalent 4×4 matrix for a system consisting of a quarter-wave plate and a stopped calcite crystal *in that order* is given by

$$\frac{1}{2}\begin{bmatrix} 1 & C_2' & S_2' & 0 \\ 1 & C_2' & S_2' & 0 \\ 0 & 0 & 0 & 0 \\ 0 & 0 & 0 & 0 \end{bmatrix}_{c''c'} \begin{bmatrix} 1 & 0 & 0 & 0 \\ 0 & C_2 & S_2 & 0 \\ 0 & 0 & 0 & -1 \\ 0 & -S_2 & C_2 & 0 \end{bmatrix}_{c'c}$$

$$= \frac{1}{2}\begin{bmatrix} 1 & C_2'C_2 & C_2'S_2 & -S_2' \\ 1 & C_2'C_2 & C_2'S_2 & -S_2' \\ 0 & 0 & 0 & 0 \\ 0 & 0 & 0 & 0 \end{bmatrix}_{c''c}, \qquad (1.40)$$

where $C_2' = \cos 2\alpha'$, $C_2 = \cos 2\alpha$, etc. Thus when an arbitrary beam is passed through this system, the resultant beam is given by

$$S_{c''} = \frac{1}{2}\begin{bmatrix} 1 & C_2'C_2 & C_2'S_2 & -S_2' \\ 1 & C_2'C_2 & C_2'S_2 & -S_2' \\ 0 & 0 & 0 & 0 \\ 0 & 0 & 0 & 0 \end{bmatrix}_{c''c} \begin{bmatrix} I^{(0)} \\ P_1 \\ P_2 \\ P_3 \end{bmatrix}_c = \begin{bmatrix} I^{(2)} \\ I^{(2)} \\ 0 \\ 0 \end{bmatrix}_{c''}, \qquad (1.41)$$

and has intensity $I^{(2)} = \frac{1}{2}(I^{(0)} + P_1 C_2' C_2 + P_2 C_2' S_2 - P_3 S_2')$. For unpolarized light it is found that $I^{(2)} = \frac{1}{2}I^{(0)}$ for all α and α' so that we must have $P_1 = P_2 = P_3 = 0$.

The Stokes vector can be treated on a phenomenological basis without reference to an underlying theory. However, it may be related to the electromagnetic theory if the four elements on the right-hand side of eqn (1.25) are considered to be time-averaged quantities. Thus for unpolarized light we have

$$S_c^{\text{unpol}} = \begin{bmatrix} I \\ 0 \\ 0 \\ 0 \end{bmatrix}_c = \begin{bmatrix} \langle a^2 + b^2 \rangle \\ \langle a^2 - b^2 \rangle \\ \langle 2ab \cos \delta \rangle \\ \langle 2ab \sin \delta \rangle \end{bmatrix}_c = \begin{bmatrix} 2\langle a \rangle^2 \\ 0 \\ 0 \\ 0 \end{bmatrix}_c, \qquad (1.42)$$

since a, b, and δ will vary erratically and on a time average (represented by angular brackets) $\langle a^2 \rangle = \langle b^2 \rangle$ and $\langle \cos \delta \rangle = \langle \sin \delta \rangle = 0$.

1.7.1. *Incoherent beams*

The light emitted from different portions of most light sources is incoherent. Two light beams are said to be 'incoherent' if the Stokes parameters for the combined beam are the sums of the corresponding Stokes parameters for the two components. Thus the polarization state of the combined beam is the matrix sum of the individual Stokes vectors (for the same axes), i.e.

$$S_c^{(c)} = S_c^{(1)} + S_c^{(2)}. \tag{1.43}$$

On the other hand, if the combined beam is described by the sum of two Jones vectors, i.e. $J_c^{(c)} = J_c^{(1)} + J_c^{(2)}$, the two beams are said to be coherent, and in general the resultant intensity is not the sum of the component intensities. The exception is when the two beams have orthogonal states of polarization; such beams do not show interference effects.

Unpolarized light therefore may be considered as an incoherent mixture of equal amounts of two oppositely polarized beams, e.g.

$$\begin{bmatrix} I \\ 0 \\ 0 \\ 0 \end{bmatrix}_c = \tfrac{1}{2} \left\{ \begin{bmatrix} I \\ I \\ 0 \\ 0 \end{bmatrix}_c + \begin{bmatrix} I \\ -I \\ 0 \\ 0 \end{bmatrix}_c \right\}, \tag{1.44}$$

and the unpolarized beam has been written as the sum of two orthogonal linearly polarized beams. Alternatively, we may write

$$\begin{bmatrix} I \\ 0 \\ 0 \\ 0 \end{bmatrix}_c = \tfrac{1}{2} \left\{ \begin{bmatrix} I \\ 0 \\ 0 \\ I \end{bmatrix}_c + \begin{bmatrix} I \\ 0 \\ 0 \\ -I \end{bmatrix}_c \right\}, \tag{1.45}$$

so that the unpolarized beam is represented as the sum of right-handed and left-handed circularly polarized beams.

1.7.2. Partially polarized light: degree of polarization D

The sum of the Stokes vectors of a completely polarized beam and an unpolarized beam gives rise to a *partially* polarized beam, i.e.

$$\begin{bmatrix} I \\ P_1 \\ P_2 \\ P_3 \end{bmatrix}_c + \begin{bmatrix} I' \\ 0 \\ 0 \\ 0 \end{bmatrix}_c = \begin{bmatrix} I+I' \\ P_1 \\ P_2 \\ P_3 \end{bmatrix}_c, \tag{1.46}$$

where $P_1^2 + P_2^2 + P_3^2 = I^2 < (I+I')^2 = (I^{(c)})^2$. The degree of polarization D is defined by

$$D = I/I^{(c)} = (P_1^2 + P_2^2 + P_3^2)^{\frac{1}{2}}/I^{(c)} < 1. \tag{1.47}$$

For a completely polarized beam $D = 1$ and for unpolarized light $D = 0$.

This method of describing unpolarized and partially polarized light allows the Poincaré sphere representation to be generalized to include all types of light. The state of polarization of a beam of light with Stokes parameters I, P_1, P_2, and P_3 is represented by a vector $\mathbf{D}$ whose components are P_1/I, P_2/I, and P_3/I. For completely polarized light, $\mathbf{D}$ has unit magnitude and all points on the sphere of unit radius represent polarized light as described in Section 1.3. Points inside the sphere with $D = |\mathbf{D}| < 1$ represent partially polarized light and the origin, where $D = 0$, corresponds to unpolarized light. Opposite states of polarization lie on opposite sides of the sphere, i.e. have vectors $\pm\mathbf{D}$.

1.7.3 Partially polarized light in the Jones calculus: the density matrix

The Jones vector does not allow us to represent states of partial polarization or the unpolarized state; a given vector $\begin{bmatrix} a \\ b\,e^{i\delta} \end{bmatrix}_c$ always describes a polarized beam. However, partially polarized and unpolarized light may be included in the Jones formalism by generalizing the Jones vector to the *density matrix*. For a polarized beam, the density matrix is

$$\rho_c = J_c J_c^{\dagger} = \begin{bmatrix} a^2 & ab\,e^{-i\delta} \\ ab\,e^{i\delta} & b^2 \end{bmatrix}_c = \begin{bmatrix} \rho_{11} & \rho_{12} \\ \rho_{21} & \rho_{22} \end{bmatrix}_c, \tag{1.48}$$

where $J_c^{\dagger}$ is the adjoint of J_c and has several properties:

(1) the diagonal elements give the relative intensities of the two orthogonal linearly polarized components;
(2) the intensity of the beam is given by the trace, i.e. $I = a^2 + b^2 = \mathrm{tr}\,\rho_c$;
(3) the off-diagonal elements contain information about the relative phase δ of the two components.

The density matrix contains the same number of elements as the Stokes vector; we now show that it contains the same information. The Stokes parameters are simple combinations of the elements of the density matrix; we have

$$\left.\begin{aligned} I &= \rho_{11} + \rho_{22} \\ P_1 &= \rho_{11} - \rho_{22} \\ P_2 &= \rho_{12} + \rho_{21} \\ P_3 &= i(\rho_{12} - \rho_{21}) \end{aligned}\right\}, \tag{1.49}$$

or conversely

$$\left.\begin{aligned} \rho_{11} &= \tfrac{1}{2}(I + P_1) \\ \rho_{22} &= \tfrac{1}{2}(I - P_1) \\ \rho_{12} &= \tfrac{1}{2}(P_2 - iP_3) \\ \rho_{21} &= \tfrac{1}{2}(P_2 + iP_3) \end{aligned}\right\}. \tag{1.50}$$

Thus, for polarized light, we have

$$\rho_c = \tfrac{1}{2} \begin{bmatrix} I+P_1 & P_2-iP_3 \\ P_2+iP_3 & I-P_1 \end{bmatrix}_c. \tag{1.51}$$

We now assume that eqn (1.51) (like the Stokes parameters) describes all kinds of light. Thus for unpolarized light ($P_1 = P_2 = P_3 = 0$) we have

$$\rho_c^{unpol} = \tfrac{1}{2} \begin{bmatrix} I & 0 \\ 0 & I \end{bmatrix}_c. \tag{1.52}$$

This form is also obtained if the elements of eqn (1.48) are considered to be time-averaged quantities, since for unpolarized light $\langle a^2 \rangle = \langle b^2 \rangle$ and $\langle ab\, e^{i\delta} \rangle = \langle ab\, e^{-i\delta} \rangle = 0$. The polarization state of a beam of light may be described by the matrix sum of density matrices (cf Stokes vectors) corresponding to any number of appropriate incoherent component beams. Thus unpolarized light may be considered to be an equal mixture of two incoherent beams of opposite polarization, namely,

$$\tfrac{1}{2}\begin{bmatrix} I & 0 \\ 0 & I \end{bmatrix}_c = \tfrac{1}{2}\left\{ \tfrac{1}{2}\begin{bmatrix} I+P_1 & P_2-iP_3 \\ P_2+iP_3 & I-P_1 \end{bmatrix}_c + \tfrac{1}{2}\begin{bmatrix} I-P_1 & -P_2+iP_3 \\ -P_2-iP_3 & I+P_1 \end{bmatrix}_c \right\}. \tag{1.53}$$

1.7.4. Interaction between light and optical devices in the density matrix formalism

We now wish to obtain the procedure for finding the density matrix of the resultant beam of light after transmission through an optical device. For polarized beams we have $J_{c'}^{(1)} = T_{c'c}J_c^{(0)}$, consequently

$$\rho_{c'}^{(1)} = J_{c'}^{(1)}J_{c'}^{(1)\dagger} = T_{c'c}J_c^{(0)}J_c^{(0)\dagger}T_{c'c}^{\dagger} = T_{c'c}\rho_c^{(0)}T_{c'c}^{\dagger} \tag{1.54}$$

gives the relationship between $\rho_{c'}^{(1)}$ and $\rho_c^{(0)}$ in terms of $T_{c'c}$. This procedure also holds for beams which are unpolarized or partially polarized. The resultant beam is described by a density matrix which is the product of *three* 2×2 matrices. Thus, while for polarized beams the Jones calculus involving 2×2 matrices operating on 2×1 matrices is simpler than the Mueller calculus which involves 4×4 matrices acting upon 4×1 matrices, the extension to the density matrix representation necessary for a complete specification of all kinds of polarization leads to rather greater complexity.

As a simple example of using the density matrix consider the transmission of linearly polarized light $J_c^{(0)} = \begin{bmatrix} a \\ 0 \end{bmatrix}_c$ by a stopped calcite crystal. We have

$$\rho_{c'}^{(1)} = \begin{bmatrix} \cos\alpha & \sin\alpha \\ 0 & 0 \end{bmatrix}_{c'c} \begin{bmatrix} a^2 & 0 \\ 0 & 0 \end{bmatrix}_c \begin{bmatrix} \cos\alpha & 0 \\ \sin\alpha & 0 \end{bmatrix}_{c'c} = \begin{bmatrix} a^2\cos^2\alpha & 0 \\ 0 & 0 \end{bmatrix}_{c'}, \tag{1.55}$$

which represents linearly polarized light of intensity $I^{(1)} = I^{(0)}\cos^2\alpha$ as in eqn (1.20).

1.8. Comparison of Jones and Mueller calculi

The two methods have their advantages and disadvantages and to some extent complement one another. The Jones calculus is suitable for dealing with coherent beams and is simpler for polarized light. The Mueller method uses observable quantities and is not dependent upon a theory such as the electromagnetic theory. Furthermore, partially polarized and unpolarized light are incorporated naturally into this formalism while to do this the Jones calculus requires an extension to the density matrix with its greater complexity. However, the Mueller method is still more general, because although the density matrix contains the same information as the Stokes vector S, the transformation matrix Z is more general than the transformation operator T. For every matrix T in the Jones calculus there is a matrix Z in the Mueller calculus but the converse is not true. For example, suppose we have an optical device which irrespective of the type of incident light produces unpolarized light, i.e. a complete depolarizer; in the Mueller calculus we could represent this device by

$$Z_{c'c}^{\text{depol}} = \begin{bmatrix} 1 & 0 & 0 & 0 \\ 0 & 0 & 0 & 0 \\ 0 & 0 & 0 & 0 \\ 0 & 0 & 0 & 0 \end{bmatrix}_{c'c}, \tag{1.56}$$

which acting upon an arbitrary Stokes vector leads to unpolarized light with intensity $I^{(1)} = I^{(0)}$. There is no equivalent $T_{c'c}$ representation of $Z_{c'c}^{\text{depol}}$ in the Jones formalism. No such cases have yet been found in nature.

1.8.1. Calculation of Z from T

We now demonstrate that for every T there is a corresponding unique Z. Writing

$$T_{c'c} = \begin{bmatrix} T_{11} & T_{12} \\ T_{21} & T_{22} \end{bmatrix}_{c'c} \tag{1.57}$$

and using eqns (1.48) and (1.54), we can write the elements of $\rho_c^{(1)}$ as a linear function of the elements of $\rho_c^{(0)}$, e.g. (denoting complex conjugation by *)

$$\rho_{11}^{(1)} = (T_{11}T_{11}^*)\rho_{11}^{(0)} + (T_{11}T_{12}^*)\rho_{12}^{(0)} + (T_{12}T_{11}^*)\rho_{21}^{(0)} + (T_{12}T_{12}^*)\rho_{22}^{(0)}. \tag{1.58}$$

These linear relations may be expressed in matrix form, namely,

$$D_{c'}^{(1)} = \begin{bmatrix} \rho_{11}^{(1)} \\ \rho_{12}^{(1)} \\ \rho_{21}^{(1)} \\ \rho_{22}^{(1)} \end{bmatrix} = \begin{bmatrix} T_{11}T_{11}^* & T_{11}T_{12}^* & T_{12}T_{11}^* & T_{12}T_{12}^* \\ T_{11}T_{21}^* & T_{11}T_{22}^* & T_{12}T_{21}^* & T_{12}T_{22}^* \\ T_{21}T_{11}^* & T_{21}T_{12}^* & T_{22}T_{11}^* & T_{22}T_{12}^* \\ T_{21}T_{21}^* & T_{21}T_{22}^* & T_{22}T_{21}^* & T_{22}T_{22}^* \end{bmatrix}_{c'c} \begin{bmatrix} \rho_{11}^{(0)} \\ \rho_{12}^{(0)} \\ \rho_{21}^{(0)} \\ \rho_{22}^{(0)} \end{bmatrix}_c = (T_{c'c} \otimes T_{c'c}^*)D_c^{(0)}, \tag{1.59}$$

where the symbol $\otimes$ denotes the 'direct product' of the matrices $T_{c'c}$ and $T_{c'c}^*$. The Stokes vector (see eqn (1.49)) is a linear transformation of the D matrix, and we have

$$S_c = UD_c, \tag{1.60}$$

where

$$U = \begin{bmatrix} 1 & 0 & 0 & 1 \\ 1 & 0 & 0 & -1 \\ 0 & 1 & 1 & 0 \\ 0 & i & -i & 0 \end{bmatrix}. \tag{1.61}$$

Thus $S_{c'}^{(1)} = UD_{c'}^{(1)} = U(T_{c'c} \otimes T_{c'c}^*)D_c^{(0)} = U(T_{c'c} \otimes T_{c'c}^*)U^{-1}S_c^{(0)}$, so that

$$Z_{c'c} = U(T_{c'c} \otimes T_{c'c}^*)U^{-1}, \tag{1.62}$$

where

$$U^{-1} = \tfrac{1}{2}\begin{bmatrix} 1 & 1 & 0 & 0 \\ 0 & 0 & 1 & -i \\ 0 & 0 & 1 & i \\ 1 & -1 & 0 & 0 \end{bmatrix}. \tag{1.63}$$

1.9. Measurement of Stokes parameters

In this section we show how the Stokes parameters for an arbitrary beam of light may be determined employing a simple set of intensity measurements.

First, we pass the light through a stopped calcite crystal. For convenience we define the polarization state of the incident beam relative to axes which coincide with the crystal axes, i.e. $\alpha = 0$ in eqn (1.33), and the resultant beam is given by (since axes c' ≡ axes c)

$$S_{c'}^{(1)} = \tfrac{1}{2}\begin{bmatrix} 1 & 1 & 0 & 0 \\ 1 & 1 & 0 & 0 \\ 0 & 0 & 0 & 0 \\ 0 & 0 & 0 & 0 \end{bmatrix}_{c'c} \begin{bmatrix} I^{(0)} \\ P_1 \\ P_2 \\ P_3 \end{bmatrix}_c = \tfrac{1}{2}\begin{bmatrix} I^{(0)}+P_1 \\ I^{(0)}+P_1 \\ 0 \\ 0 \end{bmatrix}_{c'} \equiv \tfrac{1}{2}\begin{bmatrix} I^{(0)}+P_1 \\ I^{(0)}+P_1 \\ 0 \\ 0 \end{bmatrix}_c. \tag{1.64}$$

The intensity of the transmitted radiation is

$$I^{(1)} = \tfrac{1}{2}(I^{(0)}+P_1). \tag{1.65}$$

We now rotate the crystal through 90° about the z'-axis, so that $\alpha = \tfrac{1}{2}\pi$. The transmitted beam is

$$S_{c''}^{(2)} = \tfrac{1}{2}\begin{bmatrix} 1 & -1 & 0 & 0 \\ 1 & -1 & 0 & 0 \\ 0 & 0 & 0 & 0 \\ 0 & 0 & 0 & 0 \end{bmatrix}_{c''c'} \begin{bmatrix} I^{(0)} \\ P_1 \\ P_2 \\ P_3 \end{bmatrix}_{c'} = \tfrac{1}{2}\begin{bmatrix} I^{(0)}-P_1 \\ I^{(0)}-P_1 \\ 0 \\ 0 \end{bmatrix}_{c''}, \tag{1.66}$$

so that the measured intensity is

$$I^{(2)} = \tfrac{1}{2}(I^{(0)} - P_1).$$ (1.67)

From eqns (1.65) and (1.67) we have (since intensities are independent of axes)

$$I^{(0)} = I^{(1)} + I^{(2)} \quad \text{and} \quad P_1 = I^{(1)} - I^{(2)}.$$ (1.68)

To obtain P_2 we observe the resultant light intensity with $\alpha = \tfrac{1}{4}\pi$ and $\alpha = -\tfrac{1}{4}\pi$ respectively; we have

$$S_{c\pm}^{(\pm)} = \tfrac{1}{2}
\begin{bmatrix}
1 & 0 & \pm 1 & 0 \\
1 & 0 & \pm 1 & 0 \\
0 & 0 & 0 & 0 \\
0 & 0 & 0 & 0
\end{bmatrix}_{c\pm c'}
\begin{bmatrix}
I^{(0)} \\
P_1 \\
P_2 \\
P_3
\end{bmatrix}_{c'}
= \tfrac{1}{2}
\begin{bmatrix}
I^{(0)} \pm P_2 \\
I^{(0)} \pm P_2 \\
0 \\
0
\end{bmatrix}_{c\pm},$$ (1.69)

with intensities $I^{(\pm)} = \tfrac{1}{2}(I^{(0)} \pm P_2)$ for $\alpha = \pm\tfrac{1}{4}\pi$, so that

$$P_2 = I^{(+)} - I^{(-)}.$$ (1.70)

Finally, to determine P_3, we insert a quarter-wave plate with arbitrary orientation β between the light beam and the stopped calcite crystal and observe the transmitted intensities for $\alpha = \pm\tfrac{1}{4}\pi$ again. We have

$$S_{\tilde{c}\pm}''^{(\pm)} = \tfrac{1}{2}
\begin{bmatrix}
1 & 0 & \pm 1 & 0 \\
1 & 0 & \pm 1 & 0 \\
0 & 0 & 0 & 0 \\
0 & 0 & 0 & 0
\end{bmatrix}_{\tilde{c}\pm c'''}
\begin{bmatrix}
1 & 0 & 0 & 0 \\
0 & C_2 & S_2 & 0 \\
0 & 0 & 0 & -1 \\
0 & -S_2 & C_2 & 0
\end{bmatrix}_{c'''c'}
\begin{bmatrix}
I^{(0)} \\
P_1 \\
P_2 \\
P_3
\end{bmatrix}_{c'}$$

$$= \tfrac{1}{2}
\begin{bmatrix}
I^{(0)} \mp P_3 \\
I^{(0)} \mp P_3 \\
0 \\
0
\end{bmatrix}_{\tilde{c}\pm},$$ (1.71)

where $C_2 = \cos 2\beta$ and $S_2 = \sin 2\beta$ and the intensities $I'^{(\pm)} = \tfrac{1}{2}(I^{(0)} \mp P_3)$ for $\alpha = \pm\tfrac{1}{4}\pi$, giving

$$P_3 = I'^{(-)} - I'^{(+)}.$$ (1.72)

Thus all the Stokes parameters have been measured. It should be noted that the four elements of the equivalent density matrix of the Jones formalism describing the incident beam may also be determined by the same set of measurements. Furthermore, unknown Mueller matrices $Z_{c'c}$ (or the corresponding Jones matrices $T_{c'c}$) may be determined by appropriate similar measurements employing light beams of known polarizations.

2

QUANTUM THEORY TREATMENT
OF POLARIZATION

IN the previous discussion we have neglected the fact that the energy of a light beam is quantized. We now indicate that when a light beam is considered to consist of photons instead of electromagnetic waves the formalism of Chapter 1 remains valid but the interpretation is different.

2.1. Jones calculus

Consider the passage of a linearly polarized light wave through a stopped calcite crystal, with the incident beam having a line of polarization at an angle α to that of the emergent beam. From Section 1.5.1 the Jones vector for the transmitted beam is

$$J_{c'} = \begin{bmatrix} \cos \alpha & \sin \alpha \\ 0 & 0 \end{bmatrix}_{c'c} \begin{bmatrix} a \\ 0 \end{bmatrix}_c = \begin{bmatrix} a \cos \alpha \\ 0 \end{bmatrix}_{c'}, \qquad (2.1)$$

and the resultant intensity is $\cos^2 \alpha$ times the incident beam intensity. In order to interpret the same result in the quantum theory, *each* photon is considered to have a *probability* of $\cos^2 \alpha$ of passing through the crystal. A single photon will either pass through or be stopped by the crystal but, in accordance with the *correspondence principle*, in the limit of very many photons a fraction $\cos^2 \alpha$ will be transmitted. The probability concept arises because the intensity of a beam is proportional to the number of photons per unit volume and the photons are considered to be both indivisible and indistinguishable. Thus the elements of the Jones vector are interpreted as probability amplitudes in the sense that a^2 and b^2 give the relative probabilities of a photon being found in the two states of linear polarization respectively. For many photons per unit volume, a^2 and b^2 give the relative numbers of photons in the two states which is essentially the classical interpretation since the intensity depends upon the number of photons per unit volume.

In the electromagnetic theory, completely polarized light is described in terms of two perpendicular vibrations transverse to the direction of propagation, namely,

$$\mathbf{E} = (a\mathbf{e}_x + be^{i\delta}\mathbf{e}_y) e^{i(kz - \omega t)}. \qquad (2.2)$$

Analogously, in the quantum theory the polarization state of a completely polarized beam of photons is written as a linear superposition of two opposite polarization states ϕ_1 and ϕ_2, i.e.

$$\chi = a_1\phi_1 + a_2\phi_2, \tag{2.3}$$

where the probability amplitudes a_1 and a_2 are complex numbers. In the classical theory the intensity of a beam is measured by the quantity $(a^2 + b^2)$. Similarly, the intensity of a photon beam is described by the quantity $(|a_1|^2 + |a_2|^2)$. If the polarization states are represented by two component column matrices:

$$\phi_1 = \begin{bmatrix} 1 \\ 0 \end{bmatrix}_c, \qquad \phi_2 = \begin{bmatrix} 0 \\ 1 \end{bmatrix}_c, \tag{2.4}$$

where the subscript c again denotes reference axes we have

$$\chi = \begin{bmatrix} a_1 \\ a_2 \end{bmatrix}_c. \tag{2.5}$$

Moreover, if the two basis states of eqn (2.4) are taken to be orthogonal states of linear polarization the state vector χ is essentially the Jones vector of eqn (1.11). In this representation the interaction operators $T_{c'c}$ of Chapter 1 remain unchanged and may be used to describe the interaction of a photon beam with various optical devices. In principle, however, the states ϕ_1 and ϕ_2 may be any two opposite states of polarized light, and we may with equal validity use the two states of right-handed and left-handed circularly polarized light. However, in such a representation the operators $T_{c'c}$ would require modification.

For polarized photons, a density matrix ρ may be formed in the same manner as in Section 1.7.3, i.e.

$$\rho_c = \chi_c\chi_c^\dagger = \begin{bmatrix} a_1 \\ a_2 \end{bmatrix}_c [a_1^* a_2^*]_c = \begin{bmatrix} |a_1|^2 & a_1a_2^* \\ a_1^*a_2 & |a_2|^2 \end{bmatrix}_c = \begin{bmatrix} \rho_{11} & \rho_{12} \\ \rho_{21} & \rho_{22} \end{bmatrix}_c. \tag{2.6}$$

The general density matrix also describes partially polarized and unpolarized photons and, provided one re-interprets relative intensities as relative probabilities, retains its properties: the diagonal elements represent the relative probabilities of finding a photon in the states ϕ_1 and ϕ_2, the trace of the matrix measures the total probability of finding a photon (i.e. the intensity of the beam) and the off-diagonal elements give the relative phase. Hence the Jones calculus remains basically unchanged and only the interpretation is different in the transition from classical to quantum theory.

2.2. Mueller calculus

From eqns (1.25) and (1.49) the Stokes vector is given by

$$S_c = \begin{bmatrix} I \\ P_1 \\ P_2 \\ P_3 \end{bmatrix}_c = \begin{bmatrix} \rho_{11} + \rho_{22} \\ \rho_{11} - \rho_{22} \\ \rho_{12} + \rho_{21} \\ i(\rho_{12} - \rho_{21}) \end{bmatrix}_c. \tag{2.7}$$

Thus for polarized photons

$$S_c = \begin{bmatrix} |a_1|^2 + |a_2|^2 \\ |a_1|^2 - |a_2|^2 \\ a_1 a_2^* + a_1^* a_2 \\ i(a_1 a_2^* - a_1^* a_2) \end{bmatrix}_c, \tag{2.8}$$

which is analogous to the right-hand side of eqn (1.25). The elements of the Stokes vector represent measurable quantities which are quantum-mechanical averages over an ensemble of photons (see Section 2.4.1). If we use the basis functions of eqn (2.4), the Mueller matrices $Z_{c'c}$ corresponding to each $T_{c'c}$ operator of the Jones calculus remain unchanged, since $Z_{c'c}$ is obtainable from $T_{c'c}$. Thus there is essentially no change in the formalism developed in Chapter 1.

2.3. Spin of photon

So far we have represented a beam of photons by a two-component spinor, i.e. the state vector is considered to be completely described in terms of two basis functions as in eqn (2.3). Such a representation is usually employed in a non-relativistic description of spin-$\frac{1}{2}$ particles such as electrons. However, the photon is known to have spin 1, and one may therefore expect that the photon (like the deuteron) should be represented by a three-component vector. To understand this point and to introduce the usual angular momentum operators, we consider the angular momentum associated with scalar and vector fields.

2.3.1. *Angular momentum in a scalar field*

Any function that is a single function of the space coordinates will form a scalar field, i.e. $\psi = \psi(x, y, z)$. If we consider an infinitesimal rotation $\delta\theta$ about the z-axis which carries the coordinates x, y, z into x', y', z', we may write

$$\left. \begin{array}{l} x' = x + y\,\delta\theta \\ y' = y - x\,\delta\theta \\ z' = z \end{array} \right\}, \tag{2.9}$$

where we have used the approximations $\cos\theta \simeq 1, \sin\theta \simeq \delta\theta$ for infinitesimal $\delta\theta$. Using Taylor's expansion to first order we may write

$$\psi(x', y', z') = \psi(x, y, z) - \delta\theta \left(x\frac{\partial}{\partial y} - y\frac{\partial}{\partial x} \right) \psi(x, y, z)$$

$$= (1 - i\,\delta\theta L_z/\hbar)\psi(x, y, z), \tag{2.10}$$

where we have used the well-known expression for the z-component of orbital angular momentum

$$L_z = -i\hbar\left(x\frac{\partial}{\partial y} - y\frac{\partial}{\partial x}\right) = (-i\hbar\mathbf{r}\times\mathbf{V})_z. \qquad (2.11)$$

Thus an infinitesimal rotation has an orbital angular momentum operator associated with it. Similarly for infinitesimal rotations $\delta\theta$ about the x- and y-axes we have, respectively,

$$\psi(x'', y'', z'') = (1 - i\delta\theta L_x/\hbar)\psi(x, y, z), \qquad (2.12)$$

and

$$\psi(x''', y''', z''') = (1 - i\delta\theta L_y/\hbar)\psi(x, y, z). \qquad (2.13)$$

The vector operator $\mathbf{L}$ and its components L_x, L_y, L_z satisfy the usual commutation relationships

$$\left.\begin{array}{c} L_xL_y - L_yL_x = i\hbar L_z \\ L_xL^2 - L^2L_x = 0, \quad \text{etc.} \end{array}\right\}, \qquad (2.14)$$

where

$$L^2 = L_x^2 + L_y^2 + L_z^2. \qquad (2.15)$$

The operator L^2 and one of L_x, L_y, L_z usually taken to be L_z will have simultaneous eigenvalues $l(l+1)\hbar^2$ and $m\hbar$ when operating on an eigenfunction (spherical harmonic) Y_{lm}, i.e.

$$L^2 Y_{lm} = l(l+1)\hbar^2 Y_{lm}, \qquad (2.16)$$

$$L_z Y_{lm} = m\hbar Y_{lm}, \qquad (2.17)$$

where $l = 0, 1, 2, \ldots$ and $m = l, l-1, \ldots, -l+1, -l$.

2.3.2. Angular momentum in a vector field

In a vector field (e.g. electric field $\mathbf{E}$) the quantity which is a function of x, y, and z will have both a magnitude and a direction. With each point in the field, we associate three components each of which is a function of x, y, and z. We now consider an infinitesimal rotation $\delta\theta$ about the z-axis of a vector $\mathbf{A} = (A_x, A_y, A_z)$. The new vector $\mathbf{A}'$, referred to the original axes, has components (to first order in $\delta\theta$)

$$\left.\begin{array}{l} A_x' = A_x - \delta\theta\left(x\dfrac{\partial}{\partial y} - y\dfrac{\partial}{\partial x}\right)A_x, \\[3mm] A_y' = A_y - \delta\theta\left(x\dfrac{\partial}{\partial y} - y\dfrac{\partial}{\partial x}\right)A_y, \\[3mm] A_z' = A_z - \delta\theta\left(x\dfrac{\partial}{\partial y} - y\dfrac{\partial}{\partial x}\right)A_z. \end{array}\right\} \qquad (2.18)$$

The components of $\mathbf{A}'$, referred to the rotated system x', y', and z', are (neglecting second-order terms)

$$
\left.
\begin{aligned}
A'_{x'} &= A'_x - \delta\theta A'_y = \left(1 - \frac{\mathrm{i}\delta\theta}{\hbar}L_z\right)A_x - \delta\theta A_y, \\
A'_{y'} &= A'_y + \delta\theta A'_x = \left(1 - \frac{\mathrm{i}\delta\theta}{\hbar}L_z\right)A_y + \delta\theta A_x, \\
A'_{z'} &= A'_z = \left(1 - \frac{\mathrm{i}\delta\theta}{\hbar}L_z\right)A_z
\end{aligned}
\right\}
\tag{2.19}
$$

These equations may be written in matrix form

$$
\begin{bmatrix} A'_{x'} \\ A'_{y'} \\ A'_{z'} \end{bmatrix} = \begin{bmatrix} (1-\mathrm{i}\delta\theta L_z/\hbar) & -\delta\theta & 0 \\ \delta\theta & (1-\mathrm{i}\delta\theta L_z/\hbar) & 0 \\ 0 & 0 & (1-\mathrm{i}\delta\theta L_z/\hbar) \end{bmatrix} \begin{bmatrix} A_x \\ A_y \\ A_z \end{bmatrix},
$$

i.e.

$$
A' = \left\{(1-\mathrm{i}\delta\theta L_z/\hbar)\mathbf{1}_3 - \frac{\mathrm{i}\delta\theta}{\hbar}S_z\right\}A,
\tag{2.20}
$$

where $\mathbf{1}_3$ is the 3×3 unit matrix and we have introduced the z-component of some operator $\mathbf{S}$,

$$
S_z = \hbar\begin{bmatrix} 0 & -\mathrm{i} & 0 \\ \mathrm{i} & 0 & 0 \\ 0 & 0 & 0 \end{bmatrix}.
\tag{2.21}
$$

Similarly, rotations about the x- and y-axes lead to

$$
S_x = \hbar\begin{bmatrix} 0 & 0 & 0 \\ 0 & 0 & -\mathrm{i} \\ 0 & \mathrm{i} & 0 \end{bmatrix} \quad \text{and} \quad S_y = \hbar\begin{bmatrix} 0 & 0 & \mathrm{i} \\ 0 & 0 & 0 \\ -\mathrm{i} & 0 & 0 \end{bmatrix}.
\tag{2.22}
$$

The components of $\mathbf{S}$ satisfy a set of commutation relationships similar to eqn (2.14) for $\mathbf{L}$:

$$
\left.
\begin{aligned}
S_x S_y - S_y S_x &= \mathrm{i}\hbar S_z, \\
S_x S^2 - S^2 S_x &= 0, \quad \text{etc.}
\end{aligned}
\right\}
\tag{2.23}
$$

where

$$
S^2 = S_x^2 + S_y^2 + S_z^2 = 2\hbar^2 \mathbf{1}_3.
\tag{2.24}
$$

Thus the vector field is associated with an intrinsic angular-momentum-like term $\mathbf{S}$ of magnitude $\{s(s+1)\}^{\frac{1}{2}}\hbar \equiv \sqrt{(2)}\hbar$ and the corresponding quanta of such a field are said to have spin 1 (i.e. $s = 1$). By analogy with orbital angular momentum, S_z is considered to have the three eigenvalues $m_s = 0$, $\pm \hbar$, and the corresponding normalized eigenfunctions are

$$\chi_0 = \begin{bmatrix} 0 \\ 0 \\ 1 \end{bmatrix} \quad \text{and} \quad \chi_{\pm 1} = \frac{1}{\sqrt{2}} \begin{bmatrix} 1 \\ \pm i \\ 0 \end{bmatrix}. \tag{2.25}$$

2.3.3. *Spin in the electromagnetic field*

The electromagnetic field is transverse and consequently, for waves travelling along the z-axis, $E_z \equiv 0$. In terms of photons this is equivalent to saying that the state for which S_z has eigenvalue $m_s = 0$ does not exist, i.e. a photon has zero probability of being in the spin state χ_0. In a strict sense, this means that photon beams are always partially polarized (see Section 4.1.3). However, for historical reasons and convenience, a beam of photons will be represented by the two basis vectors $\chi_{\pm 1}$, which, neglecting the third element which is always zero, are immediately recognizable as the Jones representations for right-handed and left-handed circularly polarized light, respectively. This representation of a photon beam by 2×1 matrices rather than 3×1 matrices reduces the size of the corresponding density matrices from 3×3 to 2×2 and is the essential reason why the conventional description of polarized light is not a representation in real space but is in Poincaré space.

2.4. Quantum states

It is one of the postulates of quantum mechanics that to every observable quantity there is a corresponding linear (to satisfy the superposition principle) and self-adjoint (so that the expectation values are real) operator $\hat{\Omega}$ with associated eigenfunctions ϕ_λ and eigenvalues λ given by

$$\hat{\Omega}\phi_\lambda = \lambda\phi_\lambda. \tag{2.26}$$

(We denote quantum-mechanical operators by an accent ($\hat{\ }$)). In general, the collection or spectrum of eigenvalues is a combination of both discrete and continuous values. However, the intrinsic spin angular momentum operators (e.g. $\hat{S}_z$ of eqn (2.21)) have only a discrete and finite set of eigenvalues, and we shall restrict our discussion to such operators. There are two types of quantum states, (1) pure states and (2) mixed states, associated with an operator. For example, completely polarized and partially polarized states of light are pure and mixed states, respectively.

2.4.1. *Pure states*

If a quantum state can be defined by a wave function (χ, say, where for convenience we omit the reference axes), it is said to be a *pure state*. We can expand the state χ in terms of the eigenstates ϕ_λ of some appropriate operator

$\hat{\Omega}$ which form a complete set of basis states:

$$\chi = \sum_{\lambda} a_{\lambda}\phi_{\lambda}. \tag{2.27}$$

Note that in general χ is not an eigenstate of $\hat{\Omega}$ but a coherent superposition of eigenstates. For spin spaces which have discrete, non-degenerate and finite spectra of eigenvalues, the wave function χ may be conveniently represented by single column matrices with as many rows as the possible number of eigenvalues. A single measurement of the observable will yield one of the eigenvalues λ with relative probability $|a_{\lambda}|^2$. Thus, only in the special case where χ is an eigenstate ϕ_{λ} of the operator $\hat{\Omega}$ is the result of a single observation predictable with certainty, i.e. the eigenvalue λ. The result of a series of measurements is the *expectation* or *mean value* which for a column vector χ independent of the ordinary spatial variables is given by

$$\langle \hat{\Omega} \rangle = \chi^{\dagger}\hat{\Omega}\chi, \tag{2.28}$$

where χ is assumed to be normalized, i.e. $\chi^{\dagger}\chi = 1$.

It is convenient to express the expectation value in terms of the density matrix. We have, using eqn (2.27),

$$\langle \hat{\Omega} \rangle = \sum_{\lambda\mu} a_{\lambda}^* a_{\mu}\phi_{\lambda}^{\dagger}\hat{\Omega}\phi_{\mu} = \sum_{\lambda\mu} \rho_{\mu\lambda}\Omega_{\lambda\mu}, \tag{2.29}$$

where

$$\rho_{\mu\lambda} = a_{\mu}a_{\lambda}^* \quad \text{and} \quad \Omega_{\lambda\mu} = \phi_{\lambda}^{\dagger}\hat{\Omega}\phi_{\mu}. \tag{2.30}$$

Thus

$$\langle \hat{\Omega} \rangle = \text{tr}(\rho\Omega), \tag{2.31}$$

where ρ and Ω are Hermitian matrices with elements $\rho_{\lambda\mu}$ and $\Omega_{\lambda\mu}$, respectively. In general, the matrix Ω is different from the matrix operator $\hat{\Omega}$ and the density matrix defined above cannot be expressed in the form $\chi\chi^{\dagger}$ corresponding to the definition of eqn (2.6). However, the matrices Ω and $\hat{\Omega}$ are identical and $\rho = \chi\chi^{\dagger}$ if the eigenstates ϕ_{λ} correspond to a *diagonalized* matrix operator and take the simple form

$$\phi_1 = \begin{bmatrix} 1 \\ 0 \\ 0 \\ \cdot \\ \cdot \\ \cdot \end{bmatrix}, \quad \phi_2 = \begin{bmatrix} 0 \\ 1 \\ 0 \\ \cdot \\ \cdot \\ \cdot \end{bmatrix}, \quad \text{etc.} \tag{2.32}$$

This set is usually the most convenient choice of basis states.

As an example let us consider the operator $\hat{S}_z$ of eqn (2.21) with the three eigenstates χ_{+1}, χ_0, and χ_{-1} of eqn (2.25). In this case $\hat{S}_z$ is non-diagonal and for an arbitrary normalized pure state defined by the wave function

$$\chi = a\chi_1 + b\chi_0 + c\chi_{-1}, \tag{2.33}$$

we have

$$\langle \hat{S}_z \rangle = \text{tr}(\rho S_z), \tag{2.34}$$

where

$$\rho = \begin{bmatrix} |a|^2 & ab^* & ac^* \\ a^*b & |b|^2 & bc^* \\ a^*c & b^*c & |c|^2 \end{bmatrix} \neq \chi\chi^\dagger, \tag{2.35}$$

and

$$S_z = \hbar \begin{bmatrix} 1 & 0 & 0 \\ 0 & 0 & 0 \\ 0 & 0 & -1 \end{bmatrix} \neq \hat{S}_z. \tag{2.36}$$

In principle, the ϕ_λ of eqn (2.27) are only required to be the eigenstates of any suitable complete set of operators. Thus, we may also employ the eigenstates of $\hat{S}_z$ for describing the results of measurements of the observables corresponding to the other component operators, $\hat{S}_x$ and $\hat{S}_y$ of eqn (2.22). Indeed, the expectation values of $\hat{S}_x$ and $\hat{S}_y$ are given by

$$\langle \hat{S}_x \rangle = \text{tr}(\rho S_x) \quad \text{and} \quad \langle \hat{S}_y \rangle = \text{tr}(\rho S_y), \tag{2.37}$$

where

$$S_x = \frac{\hbar}{\sqrt{2}} \begin{bmatrix} 0 & -1 & 0 \\ -1 & 0 & 1 \\ 0 & 1 & 0 \end{bmatrix} \quad \text{and} \quad S_y = \frac{\hbar}{\sqrt{2}} \begin{bmatrix} 0 & i & 0 \\ -i & 0 & -i \\ 0 & i & 0 \end{bmatrix}. \tag{2.38}$$

Furthermore, the matrices S_x, S_y, and S_z of eqns (2.36) and (2.38) also satisfy the commutation relations (2.23) and hence may be used as a representation of the spin-1 angular momentum operators. In this representation $\hat{S}_z$ is diagonal being given by eqn (2.36) and the eigenstates are

$$\chi_{+1} = \begin{bmatrix} 1 \\ 0 \\ 0 \end{bmatrix}, \quad \chi_0 = \begin{bmatrix} 0 \\ 1 \\ 0 \end{bmatrix}, \quad \text{and} \quad \chi_{-1} = \begin{bmatrix} 0 \\ 0 \\ 1 \end{bmatrix}, \tag{2.39}$$

so that $\mathbf{S} = \hat{\mathbf{S}}$. We refer to such a set of eigenstates associated with some appropriate operator (and not necessarily $\hat{S}_z$) as a *diagonalized basis*. In such a basis, the expectation value of an operator $\hat{\Omega}$, e.g. the non-diagonal form of $\hat{S}_z$ of eqn (2.21) is given by

$$\langle \hat{\Omega} \rangle = \text{tr}(\rho \Omega) = \text{tr}(\rho \hat{\Omega}) = \text{tr}(\chi \chi^\dagger \hat{\Omega}). \qquad (2.40)$$

2.4.2. Mixed states

Quantum systems also exist, e.g. partially polarized light, for which it is impossible to write down a wave function or state vector. Such states are called *mixed states* because they can be represented (although not uniquely) as incoherent superpositions of pure states described by density matrices $\rho^{(n)}$, i.e. by a density matrix

$$\rho = \sum_n w_n \rho^{(n)}, \qquad (2.41)$$

where the w_n are appropriate weighting factors. The expectation value of an operator $\hat{\Omega}$ for a mixed state is given by

$$\langle \hat{\Omega} \rangle = \text{tr}(\rho \Omega) = \text{tr}\left\{ \left(\sum_n w_n \rho^{(n)} \right) \Omega \right\}$$

$$= \sum_n w_n \, \text{tr}(\rho^{(n)} \Omega) = \sum_n w_n \langle \hat{\Omega} \rangle_n, \qquad (2.42)$$

i.e. the weighted sum of the expectation values of the operator for the pure states.

2.5. Stokes parameters and operators

The Stokes parameters are observables and therefore we can associate with I, P_1, P_2, and P_3 four operators $\hat{J}$, $\hat{\sigma}_1$, $\hat{\sigma}_2$, and $\hat{\sigma}_3$ say, respectively. In each case we can obtain the operator by using eqn (2.40) and assuming for simplicity a diagonalized basis, we have (remembering ρ is in Poincaré space)

$$I = \langle \hat{J} \rangle = \rho_{11} + \rho_{22} = \text{tr}(\rho \hat{J}), \qquad \text{therefore} \quad \hat{J} = \begin{bmatrix} 1 & 0 \\ 0 & 1 \end{bmatrix} = \hat{\mathbf{1}}_2; \quad (2.43a)$$

$$P_1 = \langle \hat{\sigma}_1 \rangle = \rho_{11} - \rho_{22} = \text{tr}(\rho \hat{\sigma}_1), \qquad \text{therefore} \quad \hat{\sigma}_1 = \begin{bmatrix} 1 & 0 \\ 0 & -1 \end{bmatrix}; \quad (2.43b)$$

$$P_2 = \langle \hat{\sigma}_2 \rangle = \rho_{12} + \rho_{21} = \text{tr}(\rho \hat{\sigma}_2), \qquad \text{therefore} \quad \hat{\sigma}_2 = \begin{bmatrix} 0 & 1 \\ 1 & 0 \end{bmatrix}; \quad (2.43c)$$

$$P_3 = \langle \hat{\sigma}_3 \rangle = i(\rho_{12} - \rho_{21}) = \text{tr}(\rho \hat{\sigma}_3), \qquad \text{therefore} \quad \hat{\sigma}_3 = \begin{bmatrix} 0 & -i \\ i & 0 \end{bmatrix}. \quad (2.43d)$$

The expressions for $\hat{\sigma}_1$, $\hat{\sigma}_2$, and $\hat{\sigma}_3$ are identical with the set of matrices which were introduced by Pauli to describe electron spin non-relativistically. However, while the corresponding Pauli spin matrices $\hat{\sigma}_z$, $\hat{\sigma}_x$, and $\hat{\sigma}_y$ refer to ordinary space, the matrix operators of eqns (2.43) are referred to the

Poincaré space in which we defined P_1, P_2, and P_3. Hence there is a strong analogy between the description of photons in Poincaré space and the non-relativistic description of spin-$\frac{1}{2}$ particles in real space (see Tolhoek and de Groot 1951).

Previously we obtained (eqn (1.51)) for a light beam

$$\rho_c = \tfrac{1}{2}\begin{bmatrix} I+P_1 & P_2-iP_3 \\ P_2+iP_3 & I-P_1 \end{bmatrix}_c, \tag{2.44}$$

which we may now write as (treating $\hat{\sigma}_1$, $\hat{\sigma}_2$, and $\hat{\sigma}_3$ as the components of a vector associated with Poincaré space)

$$\rho_c = \tfrac{1}{2}(I\hat{\mathbf{1}}_2+\mathbf{P}\cdot\hat{\boldsymbol{\sigma}})_c, \tag{2.45}$$

where

$$\mathbf{P} = \langle\hat{\boldsymbol{\sigma}}\rangle = \mathrm{tr}(\rho_c\hat{\boldsymbol{\sigma}}). \tag{2.46}$$

For further discussion on quantum states, the Stokes parameters and density matrices in the quantum theory, the reader is referred to the articles by Fano (1949, 1957).

2.5.1. Pauli spin matrices for spin-$\frac{1}{2}$ particles

The detailed structure of atomic energy levels, particularly those of the alkali elements, and other phenomena such as the Zeeman effect (see van der Waerden (1960) for historical discussion) indicated that the state of an electron cannot be described as a scalar function of the spatial coordinates alone. In order to account for this, Uhlenbeck and Goudsmit suggested that the electron behaves like a spinning top with intrinsic angular momentum $\frac{1}{2}\hbar$. This concept was described mathematically in the non-relativistic limit by Pauli, who introduced the spin operator

$$\hat{\mathbf{S}} = \tfrac{1}{2}\hbar\hat{\boldsymbol{\sigma}}, \tag{2.47}$$

with components

$$\hat{S}_x = \tfrac{1}{2}\hbar\begin{bmatrix} 0 & 1 \\ 1 & 0 \end{bmatrix}, \quad \hat{S}_y = \tfrac{1}{2}\hbar\begin{bmatrix} 0 & -i \\ i & 0 \end{bmatrix}, \quad \hat{S}_z = \tfrac{1}{2}\hbar\begin{bmatrix} 1 & 0 \\ 0 & -1 \end{bmatrix} \tag{2.48}$$

which satisfy the angular momentum commutation relations (2.23), with

$$\hat{S}^2 = \hat{S}_x^2+\hat{S}_y^2+\hat{S}_z^2 = \tfrac{3}{4}\hbar^2\hat{\mathbf{1}}_2, \tag{2.49}$$

so that $\hat{S}^2$ has eigenvalue $\frac{3}{4}\hbar^2 \equiv s(s+1)\hbar^2$ with $s = \frac{1}{2}$ and $\hat{S}_z$ has eigenvalues $\pm\frac{1}{2}\hbar$.

The normalized eigenfunctions of $\hat{S}^2$ and $\hat{S}_z$, $\chi_{\frac{1}{2}m_s}$ ($m_s = \pm\frac{1}{2}$) are

$$\chi_{\frac{1}{2}\frac{1}{2}} = \begin{bmatrix} 1 \\ 0 \end{bmatrix}_c \quad \text{and} \quad \chi_{\frac{1}{2}-\frac{1}{2}} = \begin{bmatrix} 0 \\ 1 \end{bmatrix}_c, \tag{2.50}$$

where c denotes axes in the rest frame of the electron (for relativistic effects see Chapter 7) and satisfy the relations

$$\hat{S}^2 \chi_{\frac{1}{2} m_s} = \tfrac{3}{4} \hbar^2 \chi_{\frac{1}{2} m_s} \tag{2.51}$$

and

$$\hat{S}_z \chi_{\frac{1}{2} m_s} = m_s \hbar \chi_{\frac{1}{2} m_s}. \tag{2.52}$$

In this basis, a polarized beam of electrons is described by a state vector

$$\chi_c = a \chi_{\frac{1}{2}\ \frac{1}{2}} + b \chi_{\frac{1}{2}\ -\frac{1}{2}} = \begin{bmatrix} a \\ b \end{bmatrix}_c \tag{2.53}$$

and an arbitrary beam by either a density matrix ρ_c or the equivalent of the Stokes parameters I, $\mathbf{P} = I\mathbf{p}$, where $I = \text{tr } \rho_c$ and

$$\mathbf{p} = \langle \hat{\boldsymbol{\sigma}} \rangle / I = \text{tr}(\rho_c \hat{\boldsymbol{\sigma}}) / \text{tr } \rho_c. \tag{2.54}$$

Just as all states of polarized, partially polarized, and unpolarized light may be represented by points on the surface or inside a unit sphere in Poincaré space (Section 1.7.2), i.e. by the tip of the vector $\mathbf{D}$ whose components are P_1/I, P_2/I, and P_3/I, so for electrons we may imagine a vector $\mathbf{p}$ in ordinary space such that (1) if $|\mathbf{p}| = 1$ we have a completely polarized beam, (2) if $0 < |\mathbf{p}| < 1$ we have a partially polarized beam, and (3) if $|\mathbf{p}| = 0$ the beam is unpolarized.

Writing the components of $\mathbf{p}$ for a completely polarized beam in polar form, i.e.

$$\left. \begin{array}{l} p_z = \cos \alpha \\ p_x = \sin \alpha \cos \beta \\ p_y = \sin \alpha \sin \beta \end{array} \right\}, \tag{2.55}$$

and using the relations (2.6) and (2.7) with $\mathbf{P} = I\mathbf{p}$, we can write

$$\chi_c = I^{\frac{1}{2}} \begin{bmatrix} \cos \tfrac{1}{2}\alpha\ e^{-i\beta/2} \\ \sin \tfrac{1}{2}\alpha\ e^{i\beta/2} \end{bmatrix}_c. \tag{2.56}$$

2.5.2. Stokes parameters for spin-$\frac{1}{2}$ particles

The quantities I, Ip_x, Ip_y, Ip_z for spin-$\frac{1}{2}$ particles are the equivalent of the Stokes parameters I, P_1, P_2, P_3 for light. The components of $\mathbf{p}$ along and perpendicular to the direction of particle motion are called *longitudinal* and *transverse* polarization, respectively.

Scattering experiments are generally used to determine the Stokes parameters for an arbitrary beam of spin-$\frac{1}{2}$ particles. However, as is shown in detail in the next chapter, scattering experiments are not sensitive to longitudinal polarization, which to be observed in scattering must first be transformed into transverse polarization, e.g. by using a transverse electric field. This is analogous to the detection of circularly polarized light by employing a quarter-wave plate and a stopped calcite crystal (Section 1.9). Indeed, by 'equating' single scattering with passage through a stopped calcite crystal, a transverse electric field with a quarter-wave plate, transverse polarization with linear polarization, and longitudinal polarization with circular polarization, there is a very close formal analogy between the specification and measurement of polarization for both spin-$\frac{1}{2}$ particles and light.

3

SCATTERING OF SPIN-$\frac{1}{2}$ PARTICLES

FROM the previous chapter we have seen that a completely polarized beam of spin-$\frac{1}{2}$ particles may be described in the non-relativistic limit by the spin wave function

$$\chi_c = I^{\frac{1}{2}} \begin{bmatrix} \cos \frac{1}{2}\alpha \, e^{-i\beta/2} \\ \sin \frac{1}{2}\alpha \, e^{i\beta/2} \end{bmatrix}_c, \tag{3.1}$$

where I is the beam intensity and α and β define the orientation of the polarization vector $\mathbf{p}$ (sometimes loosely referred to as the spin direction) relative to some reference axes c. This matrix χ_c is of the same form as the Jones vector J_c. In this chapter we discuss the simplest interaction process, elastic scattering by a spinless target, as a method of inducing a transformation of the polarization state of a beam of spin-$\frac{1}{2}$ particles.

3.1. Elastic scattering matrix

When a particle is elastically scattered only two things change in the centre-of-mass (c.m.) coordinate system, its direction and its spin state. As in the optical case, we describe the final spin state $\chi_{c'}^{(f)}$ as a transformation of the initial spin state $\chi_c^{(i)}$, and write

$$\chi_{c'}^{(f)} = \hat{M}_{c'c}\chi_c^{(i)}, \tag{3.2}$$

where $\hat{M}_{c'c}$ is a 2×2 matrix analogous to the operator $T_{c'c}$ of the Jones calculus and which may or may not change the reference axes. In general the matrix $\hat{M}_{c'c}$ may be written as a product of two matrices:

$$\hat{M}_{c'c} = \hat{R}_{c'c}\hat{M}_{cc}, \tag{3.3}$$

where $\hat{R}_{c'c}$ is the appropriate 2×2 matrix which changes the reference axes from c to c' and $\hat{M}_{cc}$ is the *scattering matrix* for coordinate system c.

The operators $\hat{1}_2$, $\hat{\sigma}_x$, $\hat{\sigma}_y$, and $\hat{\sigma}_z$ form a complete set of basis matrices in which any arbitrary 2×2 matrix may be expanded. Thus the general form of $\hat{M}_{cc}$ assuming rotational invariance may be written

$$\hat{M}_{cc} = g\hat{1}_2 + h_x\hat{\sigma}_x + h_y\hat{\sigma}_y + h_z\hat{\sigma}_z = g\hat{1}_2 + \mathbf{h}.\hat{\mathbf{\sigma}}, \tag{3.4}$$

where g and $\mathbf{h}$ are complex quantities which are functions of energy and angular variables. The components of $\mathbf{h}$ and $\hat{\mathbf{\sigma}}$ are usually associated with the reference axes c, which for convenience we assume are defined in the c.m. system. In order to obtain experimentally verifiable predictions we must relate $\mathbf{h}$ in some way to the physical vectors of the scattering process, i.e. the incoming and outgoing wave vectors $\mathbf{k}$ and $\mathbf{k}'$. In the c.m. system $|\mathbf{k}| = |\mathbf{k}'|$, and the three unit vectors

$$\mathbf{M} = \frac{(\mathbf{k}' - \mathbf{k})}{|\mathbf{k}' - \mathbf{k}|}, \qquad \mathbf{N} = \frac{\mathbf{k} \times \mathbf{k}'}{|\mathbf{k} \times \mathbf{k}'|}, \qquad \mathbf{K} = \frac{\mathbf{k}' + \mathbf{k}}{|\mathbf{k}' + \mathbf{k}|} \tag{3.5}$$

form a rectilinear Cartesian coordinate system. Thus taking the axes x, y, and z along the vectors $\mathbf{M}$, $\mathbf{N}$, and $\mathbf{K}$, respectively, we can write

$$\mathbf{h} \cdot \hat{\boldsymbol{\sigma}} = h_x(\mathbf{M} \cdot \hat{\boldsymbol{\sigma}}) + h_y(\mathbf{N} \cdot \hat{\boldsymbol{\sigma}}) + h_z(\mathbf{K} \cdot \hat{\boldsymbol{\sigma}}). \tag{3.6}$$

In many cases additional symmetries occur (for a general description see Emmerson (1972)). If we assume *parity conservation* (transformation gives $\mathbf{M} \to -\mathbf{M}$, $\mathbf{N} \to \mathbf{N}$, $\mathbf{K} \to -\mathbf{K}$, $\hat{\boldsymbol{\sigma}} \to \hat{\boldsymbol{\sigma}}$) and *time-reversal invariance* (transformation gives $\mathbf{M} \to \mathbf{M}$, $\mathbf{N} \to -\mathbf{N}$, $\mathbf{K} \to -\mathbf{K}$, $\hat{\boldsymbol{\sigma}} \to -\hat{\boldsymbol{\sigma}}$) for the scattering process then we require $h_x = h_z = 0$, and writing $h_y = h$, the general scattering matrix is of the form

$$\hat{M}_{cc} = g\hat{1}_2 + h(\mathbf{N} \cdot \hat{\boldsymbol{\sigma}}). \tag{3.7}$$

In this case parity conservation alone requires $h_x = h_z = 0$ so that time-reversal invariance does not place any additional constraint on the scattering matrix. If we now define our coordinate system c such that the incident beam is along the z-axis, so that $\mathbf{N}$ lies in the corresponding xy-plane, then

$$\hat{M}_{cc} = g\hat{1}_2 - h\hat{\sigma}_x \sin \phi + h\hat{\sigma}_y \cos \phi, \tag{3.8}$$

where ϕ is the azimuthal angle of the scattered beam relative to the axes c.

The scattering amplitudes g and h describe the interaction of the incident beam of particles with the target nuclei. Various models of this interaction may be employed, and the model parameters deduced by comparing theoretical predictions for measurable quantities such as the scattered intensity with experiment. In this manner the validity of the model is determined. For nucleon scattering by complex nuclei one such model is the *optical model*, which represents the interaction of a nucleon with a target nucleus by a complex potential plus a spin–orbit coupling term. This model is described in detail by Hodgson (1963). The use of an imaginary potential to allow for non-elastic processes means that the scattering matrix which describes only the elastic interaction need not satisfy time-reversal invariance. However, it is usual to insist upon *reciprocity*, i.e. that the Heisenberg scattering matrix is symmetric (Biedenharn 1959). Since it can be shown (Coester 1951) that time-reversal invariance implies reciprocity, it follows that, in the present case, parity conservation guarantees reciprocity for the scattering matrix.

3.2. Elastic scattering of neutrons by a spinless target

We now discuss the polarization of neutrons elastically scattered by a spinless target within the framework of the optical model which includes a spin–orbit interaction. The potential is assumed to be of the form

$$V(r) = V_C(r) + S(r)\hat{\boldsymbol{\sigma}} \cdot \hat{\boldsymbol{\lambda}}, \tag{3.9}$$

where $V_C(r)$ and $S(r)$ are spherically symmetric short-ranged (i.e. tend to zero faster than r^{-1} for large r) potentials which may be complex and $\hat{\boldsymbol{\lambda}} = \hat{\mathbf{L}}/\hbar$. The operators $\hat{L}_z$ and $\hat{S}_z$ $(=\frac{1}{2}\hbar\hat{\sigma}_z)$ do not commute with $\hat{\boldsymbol{\sigma}} \cdot \hat{\boldsymbol{\lambda}}$, so that the wave functions $Y_{lm}(\theta, \phi)\chi_{\frac{1}{2}m_s}$ are not eigenfunctions of a Hamiltonian ($\hat{H}$ say) which contains a spin–orbit term. Only certain linear combinations

$$\psi_{jlm_j} = \sum_m C(l\tfrac{1}{2}j, mm_sm_j)Y_{lm}\chi_{\frac{1}{2}m_s}R_{jl}(r), \tag{3.10}$$

where $C(l\tfrac{1}{2}j, mm_sm_j)$ are Clebsch–Gordan coefficients as defined by Rose (1957) and $R_{jl}(r)$ are radial wave functions, commute with $\hat{H}$. These wave functions are simultaneous eigenfunctions of $\hat{J}^2$ ($\hat{\mathbf{J}} = \hat{\mathbf{L}} + \hat{\mathbf{S}}$), $\hat{L}^2$, $\hat{S}^2$, and $\hat{J}_z = \hat{L}_z + \hat{S}_z$ with corresponding eigenvalues $j(j+1)\hbar^2$, $l(l+1)\hbar^2$, $\frac{3}{4}\hbar^2$, and $m_j = m + m_s$. More explicitly, we have

$$\psi_{jlm_j} = \begin{bmatrix} \left(\dfrac{l+m_j+\frac{1}{2}}{2l+1}\right)^{\frac{1}{2}} Y_{lm_j-\frac{1}{2}} \\[2ex] \left(\dfrac{l-m_j+\frac{1}{2}}{2l+1}\right)^{\frac{1}{2}} Y_{lm_j+\frac{1}{2}} \end{bmatrix}_c R_{l+\frac{1}{2}l} \qquad \text{for } j = l + \tfrac{1}{2} \tag{3.11a}$$

and

$$\psi_{jlm_j} = \begin{bmatrix} -\left(\dfrac{l-m_j+\frac{1}{2}}{2l+1}\right)^{\frac{1}{2}} Y_{lm_j-\frac{1}{2}} \\[2ex] \left(\dfrac{l+m_j+\frac{1}{2}}{2l+1}\right)^{\frac{1}{2}} Y_{lm_j+\frac{1}{2}} \end{bmatrix}_c R_{l-\frac{1}{2}l} \qquad \text{for } j = l - \tfrac{1}{2}, \tag{3.11b}$$

where the sign convention of Condon and Shortley (1935) for the Clebsch–Gordan coefficients has been used and the spin wave functions are represented by simple column matrices:

$$\chi_{\frac{1}{2}\,\frac{1}{2}} = \begin{bmatrix} 1 \\ 0 \end{bmatrix}_c \quad \text{and} \quad \chi_{\frac{1}{2}\,-\frac{1}{2}} = \begin{bmatrix} 0 \\ 1 \end{bmatrix}_c. \tag{3.12}$$

The radial functions R_{jl} are solutions of the equation

$$\left[\frac{d^2}{dr^2} + k^2 - \frac{2\mu}{\hbar^2}\{V_C(r) + S(r)K\} - \frac{l(l+1)}{r^2}\right]rR_{jl} = 0, \tag{3.13}$$

where μ is the reduced mass of the scattering system and K is an eigenvalue of the operator $\hat{\boldsymbol{\sigma}} \cdot \hat{\boldsymbol{\lambda}}$. To evaluate K we note that $\hat{\boldsymbol{\sigma}} \cdot \hat{\boldsymbol{\lambda}} = (\hat{J}^2 - \hat{L}^2 - \hat{S}^2)\hbar^{-2}$, and so we may write

$$(\hat{\boldsymbol{\sigma}} \cdot \hat{\boldsymbol{\lambda}})\psi_{jlm_j} = \{j(j+1) - l(l+1) - \tfrac{3}{4}\}\psi_{jlm_j} = K\psi_{jlm_j}. \tag{3.14}$$

For the two values of j corresponding to a given l, $j = l \pm \frac{1}{2}$, eqn (3.14) gives $K = l$ and $K = -l-1$, respectively. The radial functions have the asymptotic form for large r:

$$R_{l \pm \frac{1}{2}l} \simeq (kr)^{-1} \sin(kr - \tfrac{1}{2}l\pi + \eta_l^{\pm}), \tag{3.15}$$

where the $\eta_l^{\pm}$ are the phase shifts corresponding to the $(l, j = l \pm \frac{1}{2})$ partial wave.

The general solution of the elastic scattering problem for a polarized beam is a linear combination of all the eigenfunctions

$$\Psi = \sum_{jlm_j} A_{jlm_j}\psi_{jlm_j}. \tag{3.16}$$

Using eqns (3.10) and (3.11) and setting $a_{lm_j} = A_{l+\frac{1}{2}lm_j}$ and $b_{lm_j} = A_{l-\frac{1}{2}lm_j}$,

$$\Psi = \sum_{lm_j} \left\{ a_{lm_j} \left[\begin{array}{c} \left(\dfrac{l+m_j+\frac{1}{2}}{2l+1}\right)^{\frac{1}{2}} Y_{lm_j-\frac{1}{2}} \\[3mm] \left(\dfrac{l-m_j+\frac{1}{2}}{2l+1}\right)^{\frac{1}{2}} Y_{lm_j+\frac{1}{2}} \end{array} \right]_c R_{l+\frac{1}{2}l} \right.$$

$$\left. + b_{lm_j} \left[\begin{array}{c} -\left(\dfrac{l-m_j+\frac{1}{2}}{2l+1}\right)^{\frac{1}{2}} Y_{lm_j-\frac{1}{2}} \\[3mm] \left(\dfrac{l+m_j+\frac{1}{2}}{2l+1}\right)^{\frac{1}{2}} Y_{lm_j+\frac{1}{2}} \end{array} \right] R_{l-\frac{1}{2}l} \right\}. \tag{3.17}$$

For large values of r, we require that Ψ may be written as the sum of an incoming plane wave in spin state $\chi_c^{(i)}$ and an outgoing spherically scattered wave in spin state $\chi_c^{(f)} = \hat{M}_{cc}\chi_c^{(i)} \equiv \hat{M}(\theta, \phi)\chi_c^{(i)}$, so that for the incident beam along the z-axis

$$\Psi \simeq e^{ikz}\chi_c^{(i)} + r^{-1} e^{ikr} \hat{M}(\theta, \phi)\chi_c^{(i)} \quad \text{as } r \to \infty. \tag{3.18}$$

Here, for convenience, we have assumed that the $\hat{M}$ matrix leaves the reference axes unchanged, and in such cases, where possible, we shall omit the subscript c. The relation (3.18) is analogous to $\psi \simeq e^{ikz} + r^{-1} e^{ikr} f(\theta)$ for spinless particles. The form of the scattering matrix $\hat{M}(\theta, \phi)$ may be determined by considering appropriate polarized beams. First, we consider an incident unit intensity beam $\Psi_{+\frac{1}{2}}$ with spin state $\chi_{\frac{1}{2}\frac{1}{2}}$ given by eqn (3.12), i.e. $p_z = +1$ and the beam is longitudinally polarized. Asymptotically we have

$$\Psi_{+\frac{1}{2}} \simeq e^{ikz} \begin{bmatrix} 1 \\ 0 \end{bmatrix} + \frac{e^{ikr}}{r} \begin{bmatrix} M_{11} & M_{12} \\ M_{21} & M_{22} \end{bmatrix} \begin{bmatrix} 1 \\ 0 \end{bmatrix}. \tag{3.19}$$

Comparing eqns (3.17) and (3.19) we obtain the relations

$$\sum_{lm_j} \left\{ a_{lm_j} \left(\frac{l+m_j+\frac{1}{2}}{2l+1} \right)^{\frac{1}{2}} Y_{lm_j-\frac{1}{2}} R_{l+\frac{1}{2}l} - b_{lm_j} \left(\frac{l-m_j+\frac{1}{2}}{2l+1} \right)^{\frac{1}{2}} Y_{lm_j-\frac{1}{2}} R_{l-\frac{1}{2}l} \right\}$$
$$\simeq e^{ikz} + r^{-1} e^{ikr} M_{11}, \tag{3.20a}$$

and

$$\sum_{lm_j} \left\{ a_{lm_j} \left(\frac{l-m_j+\frac{1}{2}}{2l+1} \right)^{\frac{1}{2}} Y_{lm_j+\frac{1}{2}} R_{l+\frac{1}{2}l} + b_{lm_j} \left(\frac{l+m_j+\frac{1}{2}}{2l+1} \right)^{\frac{1}{2}} Y_{lm_j+\frac{1}{2}} R_{l-\frac{1}{2}l} \right\}$$
$$\simeq r^{-1} e^{ikr} M_{21}. \tag{3.20b}$$

Expanding e^{ikz} in partial waves we have

$$e^{ikz} = \sum_{lm} 4\pi i^l j_l(kr) Y^*_{lm}(\theta_k, \phi_k) Y_{lm}(\theta, \phi), \tag{3.21}$$

where $j_l(kr)$ is a spherical Bessel function and $Y^*_{lm}(\theta_k, \phi_k)$ and $Y_{lm}(\theta, \phi)$ are spherical harmonics associated with the vectors $\mathbf{k}$ and $\mathbf{r}$, respectively (for convenience, we generally omit the arguments θ, ϕ corresponding to $\mathbf{r}$). For a polarized beam with $p_z = +1$, $m_j = m+\frac{1}{2}$, so that eqn (3.21) becomes (remembering that $\mathbf{k}$ is along the z-axis)

$$e^{ikz} = \sum_{lm_j} 4\pi i^l j_l(kr) Y^*_{lm_j-\frac{1}{2}}(0,0) Y_{lm_j-\frac{1}{2}}. \tag{3.22}$$

Substituting this relation for e^{ikz} into eqns (3.20) and employing the asymptotic form of R_{jl}, we obtain four equations for each set of l, m_j by equating the coefficients of $e^{\pm ikr}$ separately, and we may solve for $a_{lm_j}, b_{lm_j}, M_{11}$, and M_{21}. In this manner we obtain

$$M_{11} = \frac{1}{2ik} \sum_l \{(l+1) \exp(2i\eta_l^+) + l \exp(2i\eta_l^-) - (2l+1)\} P_l^0(\cos\theta) \tag{3.23}$$

and

$$M_{21} = \frac{1}{2ik} \sum_l \{\exp(2i\eta_l^+) - \exp(2i\eta_l^-)\} P_l^1(\cos\theta) e^{i\phi}, \tag{3.24}$$

where the $P_l^m(\cos\theta)$ are associated Legendre functions.

Similarly, by considering an incident unit intensity beam with spin state $\chi_{\frac{1}{2}-\frac{1}{2}}$ as given by eqn (3.12), i.e. $p_z = -1$, we find

$$M_{12} = -M_{21} e^{-2i\phi} \quad \text{and} \quad M_{22} = M_{11}. \tag{3.25}$$

If we write $g = M_{11}$ and $h = iM_{12} e^{i\phi}$, so that g and h are independent of ϕ, we have

$$\hat{M}(\theta, \phi) = \begin{bmatrix} g(\theta) & -ih(\theta) e^{-i\phi} \\ ih(\theta) e^{i\phi} & g(\theta) \end{bmatrix} = g\hat{\mathbf{1}}_2 - h\hat{\sigma}_x \sin\phi + h\hat{\sigma}_y \cos\phi, \tag{3.26}$$

which is identical with eqn (3.8). If there is no spin–orbit term we can write $\eta_l^+ = \eta_l^- = \eta_l^{(0)}$ and, from eqn (3.24), $h = 0$ so that the scattering matrix is diagonal, i.e. $M = g^{(0)}\hat{1}_2$, where

$$g^{(0)} = \frac{1}{2ik}\sum_l (2l+1)\{\exp(2i\eta_l^{(0)})-1\}P_l^0(\cos\theta).\qquad(3.27)$$

3.2.1. Scattering of an unpolarized beam

For an unpolarized beam of unit intensity the density matrix is (from eqn (1.52))

$$\rho_{\text{unpol}} = \begin{bmatrix} \frac{1}{2} & 0 \\ 0 & \frac{1}{2} \end{bmatrix},\qquad(3.28)$$

and the scattered beam is described by a density matrix (analogous to eqn (1.54))

$$\begin{aligned}
\rho_0^{(f)} &= \hat{M}\rho_{\text{unpol}}\hat{M}^\dagger = \tfrac{1}{2}\hat{M}\hat{M}^\dagger \\
&= \frac{1}{2}\begin{bmatrix} g & -ih\,e^{-i\phi} \\ ih\,e^{i\phi} & g \end{bmatrix}\begin{bmatrix} g^* & -ih^*\,e^{-i\phi} \\ ih^*\,e^{i\phi} & g^* \end{bmatrix} \\
&= \frac{1}{2}\begin{bmatrix} |g|^2+|h|^2 & -2i\,\text{Re}(g^*h)\,e^{-i\phi} \\ 2i\,\text{Re}(g^*h)\,e^{i\phi} & |g|^2+|h|^2 \end{bmatrix} \\
&= \tfrac{1}{2}I_0\begin{bmatrix} 1 & -iv\,e^{-i\phi} \\ iv\,e^{i\phi} & 1 \end{bmatrix},\qquad(3.29)
\end{aligned}$$

where

$$I_0 = |g|^2+|h|^2\qquad(3.30)$$

and

$$v = 2\,\text{Re}(g^*h)/I_0,\qquad(3.31)$$

which we call the *vector scattering parameter* for the particular reaction. Comparing eqn (3.29) with eqn (2.44) we see that the scattered beam has intensity I_0 and polarization $\mathbf{p}_0 = v\mathbf{N}$ *perpendicular to the scattering plane*. Thus the elements of the scattering matrix are directly related to the intensity and polarization of the outgoing beam resulting from the scattering of an unpolarized beam of unit intensity. Indeed the nature of the scattering amplitudes can be determined by intensity observations designed to measure such quantities as I_0 and v. We discuss this further in Section 3.6.

3.3. Polarization of a beam from a polarized ion source

A beam of spin-$\frac{1}{2}$ particles from a polarized ion source (e.g. protons produced by ionization of polarized hydrogen atoms in a homogeneous magnetic field) has the polarization vector **p** along the field direction which is an axis of symmetry. From Section 2.5.1 the density matrix for the beam in terms of the Stokes parameters is

$$\rho_c = \tfrac{1}{2}I\begin{bmatrix} 1+p_z & p_x-ip_y \\ p_x+ip_y & 1-p_z \end{bmatrix}_c. \tag{3.32}$$

Thus for the z-axis along the symmetry axis we may write

$$\rho_c = N_+\rho_c^{(+)} + N_-\rho_c^{(-)} = \begin{bmatrix} N_+ & 0 \\ 0 & N_- \end{bmatrix}_c, \tag{3.33}$$

where N_+ and N_- are the relative numbers of particles in the two pure states

$$\rho_c^{(+)} = \begin{bmatrix} 1 & 0 \\ 0 & 0 \end{bmatrix}_c \quad \text{and} \quad \rho_c^{(-)} = \begin{bmatrix} 0 & 0 \\ 0 & 1 \end{bmatrix}_c \tag{3.34}$$

corresponding to unit-intensity completely polarized beams with spin 'up' and spin 'down' (i.e. $p_x = p_y = 0$, $p_z = \pm 1$), respectively. The beam has intensity and polarization given by

$$I = N_+ + N_-, \qquad p_x = p_y = 0, \quad \text{and} \quad Ip_z = N_+ - N_-. \tag{3.35}$$

It is often more convenient to refer the beam polarization to a different quantization axis (e.g. the direction of motion of the beam after extraction from the source). If this direction has polar angle α and azimuthal angle β relative to the symmetry axis, the density matrix for the new axis of quantization is (using eqn (3.1) and $\rho = \chi\chi^\dagger$ for pure states)

$$\begin{aligned}
\rho_{c'} &= \tfrac{1}{2}N_+\begin{bmatrix} (1+\cos\alpha) & \sin\alpha\,e^{-i\beta} \\ \sin\alpha\,e^{i\beta} & (1-\cos\alpha) \end{bmatrix}_{c'} + \tfrac{1}{2}N_-\begin{bmatrix} (1-\cos\alpha) & -\sin\alpha\,e^{-i\beta} \\ -\sin\alpha\,e^{i\beta} & (1+\cos\alpha) \end{bmatrix}_{c'} \\
&= \tfrac{1}{2}\begin{bmatrix} (N_++N_-)+(N_+-N_-)\cos\alpha & (N_+-N_-)\sin\alpha\,e^{i\beta} \\ (N_+-N_-)\sin\alpha\,e^{i\beta} & (N_++N_-)-(N_+-N_-)\cos\alpha \end{bmatrix}_{c'},
\end{aligned} \tag{3.36}$$

and the beam has intensity and polarization

$$\begin{aligned}
I &= N_+ + N_-, \\
Ip_x &= (N_+-N_-)\sin\alpha\cos\beta, \\
Ip_y &= (N_+-N_-)\sin\alpha\sin\beta, \\
Ip_z &= (N_+-N_-)\cos\alpha.
\end{aligned} \tag{3.37}$$

3.4. Measurement of vector scattering parameter

The vector scattering parameter may be measured by scattering a *polarized* beam which has been obtained either from a polarized ion source or by the scattering of an unpolarized beam. We now describe the latter technique (the double-scattering method) in the density matrix formalism.

First scattering. The density matrix of the scattered beam is given by eqn (3.29). If we choose the scattering to be in the xz-plane with $\phi = 0$ we have

$$\rho^{(1)} = \tfrac{1}{2}I_0^{(1)}\begin{bmatrix} 1 & -iv^{(1)} \\ iv^{(1)} & 1 \end{bmatrix} = \tfrac{1}{2}I_0^{(1)}(\hat{\mathbf{1}}_2 + v^{(1)}\hat{\sigma}_y), \qquad (3.38)$$

where $I_0^{(1)} = |g_1|^2 + |h_1|^2$, $v^{(1)} = 2\operatorname{Re}(g_1^* h_1)/I_0^{(1)}$, and the superscript or subscript (1) denotes first scattering. The intensity of the scattered beam is given by

$$\operatorname{tr}\rho^{(1)} = I_0^{(1)} = |g_1|^2 + |h_1|^2. \qquad (3.39)$$

Thus after the first scattering there is axial symmetry but the beam is no longer unpolarized. The polarization of the scattered beam may be obtained either by direct comparison of eqns (3.38) and (3.32) or by using the definition (see Section 2.5.1)

$$\mathbf{p} = \langle\hat{\boldsymbol{\sigma}}\rangle/I = \operatorname{tr}(\rho\hat{\boldsymbol{\sigma}})/\operatorname{tr}\rho. \qquad (3.40)$$

We find $p_x^{(1)} = p_z^{(1)} = 0$ and

$$p_y^{(1)} = v^{(1)} = 2\operatorname{Re}(g_1^* h_1)/I_0^{(1)}, \qquad (3.41)$$

where $v^{(1)}$ is the vector scattering parameter associated with the first elastic scattering process. Usually $|v^{(1)}| < 1$, so that the scattered beam is only partially polarized.

Second scattering. The matrix $\rho_c^{(1)}$ of eqn (3.38) is defined for the special axes (c) with the z-axis along the incident beam direction and the y-axis perpendicular to the scattering plane of the first scattering. In order that we may apply the same method to the second scattering, it is necessary to re-define our axes so that the new z'-axis is in the direction of the emergent beam from the first scattering in a laboratory (lab.) system of coordinates (c'). The simplest transformation (see Fig. 3.1) involves a rotation of $(\theta_1)_{\text{lab}}$ about the y-axis, so that x' and z' remain in the plane of the scattering. In the transformation c.m. to lab. the scattered intensity changes by the ratio of the solid angles of detection $d\Omega$, and we write $I_0 \to X$, where

$$X = I_0 \, d\Omega_{\text{cm}}/d\Omega_{\text{lab}}. \qquad (3.42)$$

The magnitude of the polarization vector is invariant, although its components will change under a rotation of axes according to the usual relations. Since the polarization vector of the once scattered beam is along the y-axis, it is unaffected by a rotation about the y-axis. Finally it should be noted that, although we are considering elastic scattering, the energy of the scattered

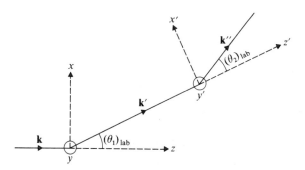

FIG. 3.1. Double scattering experiment in lab. frame. For simplicity the y- and y'-axes are taken parallel to $\mathbf{k} \times \mathbf{k}'$ and $\mathbf{k}' \times \mathbf{k}''$, respectively.

beam in the lab. frame is not the same as the incident energy since the emergent particles are degraded owing to the recoil of the target. Thus the density matrix $\rho_c^{(1)}$ becomes in the new reference frame

$$\rho_{c'}^{(1)} = (\hat{R}_{c'c}\rho_c^{(1)}\hat{R}_{c'c}^{\dagger})X^{(1)}/I_0^{(1)} = (\hat{R}_{c'c}\rho_c^{(1)}\hat{R}_{c'c}^{-1})X^{(1)}/I_0^{(1)}, \tag{3.43}$$

where

$$\hat{R}_{c'c} = \begin{bmatrix} \cos\frac{1}{2}(\theta_1)_{\text{lab}} & \sin\frac{1}{2}(\theta_1)_{\text{lab}} \\ -\sin\frac{1}{2}(\theta_1)_{\text{lab}} & \cos\frac{1}{2}(\theta_1)_{\text{lab}} \end{bmatrix}_{c'c} \tag{3.44}$$

describes the rotation $(\theta_1)_{\text{lab}}$ about the y-axis. The occurrence of half-angles in the rotation operator $\hat{R}_{c'c}$ is characteristic of the two-dimensional spin representation. One can verify readily that $\hat{R}_{c'c}$ acting upon the spin wave function χ_c of eqn (3.1) yields a state $\chi_{c'}$ whose polarization components have been modified in the required manner:

$$p_{z'} = p_z \cos(\theta_1)_{\text{lab}} + p_x \sin(\theta_1)_{\text{lab}},$$

$$p_{x'} = -p_z \sin(\theta_1)_{\text{lab}} + p_x \cos(\theta_1)_{\text{lab}},$$

$$p_{y'} = p_y. \tag{3.45}$$

Thus, since $\hat{\sigma}_{y'} = \hat{R}_{c'c}\hat{\sigma}_y\hat{R}_{c'c}^{\dagger} = \hat{\sigma}_y$,

$$\rho_{c'}^{(1)} = \tfrac{1}{2}X^{(1)}(\hat{1}_2 + v^{(1)}\hat{\sigma}_y)_{c'}. \tag{3.46}$$

Since $\rho_{c'}^{(1)}$ is unchanged when we transform from the lab. frame to the c.m. frame of the second scattering system, the density matrix for the twice-scattered beam is given (analogous to eqn (3.29)) by

$$\rho_{c'}^{(2)} = \hat{M}_{c'c}^{(2)}\rho_{c'}^{(1)}\hat{M}_{c'c}^{(2)\dagger}, \tag{3.47}$$

where c' now denotes the c.m. system. For arbitrary scattering angles θ_2 and

ϕ_2 we have

$$\rho_{c'}^{(2)} = \tfrac{1}{2}X^{(1)}\begin{bmatrix} g_2 & -ih_2\,e^{-i\phi_2} \\ ih_2\,e^{i\phi_2} & g_2 \end{bmatrix}\begin{bmatrix} 1 & -iv^{(1)} \\ iv^{(1)} & 1 \end{bmatrix}\begin{bmatrix} g_2^* & -ih_2^*\,e^{-i\phi_2} \\ ih_2^*\,e^{i\phi_2} & g_2^* \end{bmatrix}$$

$$= \tfrac{1}{2}X^{(1)}\begin{bmatrix} |g_2|^2+|h_2|^2+v^{(1)}(g_2 h_2^*\,e^{i\phi_2}+g_2^* h_2\,e^{-i\phi_2})\cdots \\ \cdots\,|g_2|^2+|h_2|^2+v^{(1)}(g_2 h_2^*\,e^{-i\phi_2}+g_2^* h_2\,e^{i\phi_2}) \end{bmatrix}. \tag{3.48}$$

Thus the intensity of the beam after the second scattering is given by

$$I_{c'}^{(2)} = \mathrm{tr}\,\rho_{c'}^{(2)} = X^{(1)}I_0^{(2)}(1+v^{(1)}v^{(2)}\cos\phi_2), \tag{3.49}$$

where $I_0^{(2)} = |g_2|^2+|h_2|^2$ and $v^{(2)} = 2\,\mathrm{Re}(g_2^* h_2)/I_0^{(2)}$ are defined similarly to $I_0^{(1)}$ and $v^{(1)}$.

The quantities $v^{(q)}$ may be considered to be the magnitudes of vectors $\mathbf{v}^{(q)} = v^{(q)}\mathbf{N}^{(q)}$, where $\mathbf{N}^{(q)}$ are unit vectors parallel to the $y^{(q)}$-axes which are along $\mathbf{k}^{(q)}\times\mathbf{k}'^{(q)}$, where $\mathbf{k}^{(q)}$ and $\mathbf{k}'^{(q)}$ are the incident and final momenta for the reaction (q) in question. Thus we can write

$$I_{c'}^{(2)} = X^{(1)}I_0^{(2)}(1+v^{(1)}v^{(2)}\mathbf{N}^{(1)}\cdot\mathbf{N}^{(2)}) = X^{(1)}I_0^{(2)}(1+\mathbf{v}^{(1)}\cdot\mathbf{v}^{(2)}). \tag{3.50}$$

Asymmetry parameter $\varepsilon_2(\theta_1,\theta_2)$. If the intensity $I_{c'}^{(2)}$ is measured for $\phi_2 = 0$ and π for given θ_1 and θ_2, the asymmetry parameter

$$\varepsilon_2(\theta_1,\theta_2) = \frac{\{I^{(2)}(\phi_2=0)-I^{(2)}(\phi_2=\pi)\}}{\{I^{(2)}(\phi_2=0)+I^{(2)}(\phi_2=\pi)\}} = v^{(1)}(\theta_1)v^{(2)}(\theta_2) \tag{3.51}$$

may be obtained, and provided one of $v^{(1)}$, $v^{(2)}$ is known the other may be determined.

3.5. Standard scattering matrix

In the previous section we have seen the necessity to keep changing the reference axes, but unless one is very careful one is liable to lose track of which reference frame one is dealing with at a given stage. It is convenient, therefore, to build all the necessary coordinate transformations into the scattering matrix. We define a 'standard' scattering matrix $\tilde{M}_{c'c}^{(q)}$ for a given elastic scattering process (q) which, for an initial density matrix ρ_c which is defined relative to some lab. axes (c) in which the z-axis is along the direction of motion of the (incident) beam of particles (see Fig. 3.2), leads to a final density matrix

$$\rho_{c'}^{(q)} = \tilde{M}_{c'c}^{(q)}\rho_c\tilde{M}_{c'c}^{(q)\dagger}, \tag{3.52}$$

which is referred to a lab. frame (c') with z'-axis along the direction of the outgoing particles and y'-axis perpendicular to the scattering plane in the

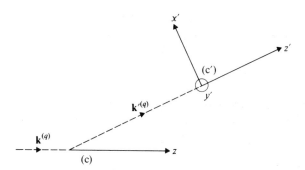

FIG. 3.2. The standard axes c,c' for reaction (q). Axis z is along $\mathbf{k}^{(q)}$. Axis z' is along $\mathbf{k}'^{(q)}$ and the y'-axis is parallel to $\mathbf{k}^{(q)} \times \mathbf{k}'^{(q)}$ in lab. frame (n.b. axes x and y (not shown) are not generally in and perpendicular to the scattering plane, respectively).

usual sense (i.e. parallel to a unit vector $\mathbf{N}$ as defined by eqn (3.5)). We have

$$
\begin{aligned}
\tilde{M}_{c'c}^{(q)} &= (X^{(q)}/I_0^{(q)})^{\frac{1}{2}} \hat{R}[(\theta_q)_{\text{lab}}] \hat{R}[\phi_q] \hat{M}(\theta_q, \phi_q) \\
&= (X^{(q)}/I_0^{(q)})^{\frac{1}{2}} \begin{bmatrix} \cos \frac{1}{2}(\theta_q)_{\text{lab}} & \sin \frac{1}{2}(\theta_q)_{\text{lab}} \\ -\sin \frac{1}{2}(\theta_q)_{\text{lab}} & \cos \frac{1}{2}(\theta_q)_{\text{lab}} \end{bmatrix} \begin{bmatrix} e^{i\phi_q/2} & 0 \\ 0 & e^{-i\phi_q/2} \end{bmatrix} \times \\
&\quad \times \begin{bmatrix} g_q & -ih_q e^{-i\phi_q} \\ ih_q e^{i\phi_q} & g_q \end{bmatrix} \\
&= (X^{(q)}/I_0^{(q)})^{\frac{1}{2}} \begin{bmatrix} G_q e^{i\phi_q/2} & -iH_q e^{-i\phi_q/2} \\ iH_q e^{i\phi_q/2} & G_q e^{-i\phi_q/2} \end{bmatrix},
\end{aligned}
\tag{3.53}
$$

where $\hat{R}[(\theta_q)_{\text{lab}}]$ (equivalent to $\hat{R}_{c'c}$ of eqn (3.44)) and $\hat{R}[\phi_q]$ are the requisite rotation matrices (in 2×2 space) to transform axes c to c'. $(X^{(q)}/I_0^{(q)})$ is the ratio of the solid angles as given by eqn (3.42) and

$$
G_q = g_q(\theta_q) \cos \frac{1}{2}(\theta_q)_{\text{lab}} + ih_q(\theta_q) \sin \frac{1}{2}(\theta_q)_{\text{lab}}
\tag{3.54}
$$

and

$$
H_q = h_q(\theta_q) \cos \frac{1}{2}(\theta_q)_{\text{lab}} + ig_q(\theta_q) \sin \frac{1}{2}(\theta_q)_{\text{lab}}.
\tag{3.55}
$$

The amplitudes G_q and H_q conveniently satisfy the relations

$$
|G_q|^2 + |H_q|^2 = |g_q|^2 + |h_q|^2 = I_0^{(q)}
\tag{3.56}
$$

and

$$
2 \operatorname{Re}(G_q^* H_q)/I_0^{(q)} = 2 \operatorname{Re}(g_q^* h_q)/I_0^{(q)} = v^{(q)}.
\tag{3.57}
$$

The results of the previous section may be obtained more easily and concisely by employing the standard scattering matrix $\tilde{M}_{c'c}^{(q)}$ of eqn (3.53).

3.5.1. *Elastic scattering of charged spin-$\frac{1}{2}$ particles*

The standard scattering matrix is again given by eqn (3.53), i.e.

$$\tilde{M}_{c'c}^{(q)} = (X^{(q)}/I_0^{(q)})^{\frac{1}{2}} \begin{bmatrix} G_q\, e^{i\phi_q/2} & -iH_q\, e^{-i\phi_q/2} \\ iH_q\, e^{i\phi_q/2} & G_q\, e^{-i\phi_q/2} \end{bmatrix}_{c'c},$$

where G_q and H_q are as defined by eqns (3.54) and (3.55). However, the amplitudes g_q and h_q now contain terms introduced by the Coulomb field and are given by (omitting the subscript q)

$$g = \frac{1}{2ik} \sum_l \{(l+1)\exp(2i\eta_l^+) + l\exp(2i\eta_l^-) - (2l+1)\} P_l^0(\cos\theta)\, e^{2i\sigma_l}$$

$$- \frac{\eta}{2k} \operatorname{cosec}^2 \frac{\theta}{2} \exp\{-i\eta\ln(\sin^2\tfrac{1}{2}\theta) + 2i\sigma_0\} \tag{3.58}$$

and

$$h = \frac{1}{2k} \sum_l \{\exp(2i\eta_l^+) - \exp(2i\eta_l^-)\} P_l^1(\cos\theta)\, e^{2i\sigma_l}, \tag{3.59}$$

where σ_l is the Coulomb phase shift given by

$$\sigma_l = \arg\Gamma(l+1+i\eta) \tag{3.60}$$

and

$$\eta = ZZ'e^2\mu/\hbar^2 k \tag{3.61}$$

is the usual Coulomb parameter for charges Ze, $Z'e$ and reduced mass μ.

3.6. Measurable quantities of the scattering matrix

The standard matrix $\tilde{M}_{c'c}$ is defined in terms of the two complex amplitudes G and H. Thus, in principle, for a given bombarding energy and for each set of scattering angles we require to measure four quantities in order to determine $\tilde{M}_{c'c}$. However, in practice, it is only possible to obtain three independent measurable quantities which enable us to obtain $\tilde{M}_{c'c}$ to within an over-all phase factor. Three such quantities are

(1) the intensity $I_0 = |G|^2 + |H|^2$, $\tag{3.62a}$

(2) the vector scattering parameter $v = 2\operatorname{Re}(G^*H)/I_0$, $\tag{3.62b}$

and

(3) the Wolfenstein parameter $W = (|G|^2 - |H|^2)/I_0$. $\tag{3.62c}$

These quantities are real and measurable; the last observable was introduced by Wolfenstein (1954) and is often denoted by R. A fourth measurable quantity may be defined:

$$w = -i(G^*H - H^*G)/I_0,$$ (3.62d)

which is the second Wolfenstein parameter often represented by A. However, w is related to the other quantities by the equation

$$W^2 + w^2 + v^2 = 1.$$ (3.63)

From the last equation we can write (regarding W, w, and v as the Cartesian components of a unit vector)

$$W = \sin \tilde{\alpha} \cos \tilde{\beta},$$ (3.64a)

$$w = \sin \tilde{\alpha} \sin \tilde{\beta},$$ (3.64b)

$$v = \cos \tilde{\alpha}.$$ (3.64c)

The ratio $w/W = \tan \tilde{\beta}$ is called the *rotation parameter*, for reasons we shall discuss shortly. In the literature a different rotation parameter is often used

$$\tan \beta' = -i(g^*h - h^*g)/(|g|^2 - |h|^2) = \tan(\tilde{\beta} - \theta_{\text{lab}}),$$ (3.65)

so that β' is simply related to $\tilde{\beta}$. For an unpolarized beam, the quantities I_0, v, and W may be obtained by a single-, double-, or triple-scattering experiment, respectively. The rather complicated triple-scattering measurements are more conveniently described using a Mueller-type formalism.

3.7. Mueller method

In the Mueller method, by analogy with the optical case, a beam of spin-$\frac{1}{2}$ particles of intensity I and polarization $\mathbf{p}$ is represented by a Stokes vector

$$S_c = I \begin{bmatrix} 1 \\ p_x \\ p_y \\ p_z \end{bmatrix}_c,$$ (3.66)

where the subscript c denotes the reference axes. To be consistent with earlier sections we assume that the z-axis is along the direction of motion of the beam, so that p_x and p_y describe the preferences for the transverse spin states along the x- and y-axes and p_z gives the preference for the longitudinal spin state along the z-axis. As in the optical case, the effect of an elastic scattering interaction with a spinless target is represented by a 4×4 matrix $Z_{c'c}$ operating upon the Stokes vector describing the incident beam to give a resultant Stokes vector for the outgoing beam, i.e. $S_c^{(f)} = Z_{c'c} S_c^{(i)}$. In Section 1.8.1 we showed how to construct the Z matrix from a given T matrix.

In a similar manner, we may obtain from the standard scattering matrix $\tilde{M}_{c'c}^{(q)}$ of eqn (3.53) the equivalent Mueller-type 4×4 $\tilde{Z}_{c'c}^{(q)}$ matrix which describes the operation of elastic scattering on the Stokes parameters. We find

$$
\tilde{Z}_{c'c}^{(q)} = X^{(q)} \begin{bmatrix} 1 & -v^{(q)} \sin \phi_q & v^{(q)} \cos \phi_q & 0 \\ 0 & W^{(q)} \cos \phi_q & W^{(q)} \sin \phi_q & -w^{(q)} \\ v^{(q)} & -\sin \phi_q & \cos \phi_q & 0 \\ 0 & w^{(q)} \cos \phi_q & w^{(q)} \sin \phi_q & W^{(q)} \end{bmatrix}_{c'c} , \tag{3.67}
$$

where the reference axes are as defined for eqn (3.53).

3.7.1. Physical significance of $\tilde{\alpha}$ and $\tilde{\beta}$

Consider a unit intensity beam of polarization $\mathbf{p}$ which is elastically scattered with $\phi = 0$ so that the scattering occurs in the xz-plane of the frame c. The resultant beam is given by

$$
X \begin{bmatrix} 1 & 0 & v & 0 \\ 0 & W & 0 & -w \\ v & 0 & 1 & 0 \\ 0 & w & 0 & W \end{bmatrix}_{c'c} \begin{bmatrix} 1 \\ p_x \\ p_y \\ p_z \end{bmatrix}_c = X \begin{bmatrix} 1+vp_y \\ Wp_x - wp_z \\ v+p_y \\ wp_x + Wp_z \end{bmatrix}_{c'}
$$

$$
= X(1+vp_y) \begin{bmatrix} 1 \\ p'_{x'} \\ p'_{y'} \\ p'_{z'} \end{bmatrix}_{c'} , \tag{3.68}
$$

where, substituting the eqns (3.64), the new polarization components are

$$
p'_{x'} = \frac{\sin \tilde{\alpha}}{(1+p_y \cos \tilde{\alpha})} (p_x \cos \tilde{\beta} - p_z \sin \tilde{\beta}), \tag{3.69a}
$$

$$
p'_{y'} = \frac{\cos \tilde{\alpha} + p_y}{(1+p_y \cos \tilde{\alpha})}, \tag{3.69b}
$$

and

$$
p'_{z'} = \frac{\sin \tilde{\alpha}}{(1+p_y \cos \tilde{\alpha})} (p_x \sin \tilde{\beta} + p_z \cos \tilde{\beta}). \tag{3.69c}
$$

Note the forms of $p'_{x'}$ and $p'_{z'}$. Except for a common factor, these expressions have the form of a rotation through an angle $\tilde{\beta}$ in the xz-plane. Consequently $\tilde{\beta}$ is a measure of the rotation of the projection of the polarization vector in the scattering plane. (Note: the actual angle of rotation is $\beta' = \tilde{\beta} - \theta_{lab}$; $p'_{x'}$ and $p'_{z'}$ are referred to the different axes c'.) The parameter $\tilde{\alpha}$ describes the rotation of the polarization vector towards the normal to the scattering plane.

3.8. Measurement of the Wolfenstein parameters W and w by triple scattering

To exhibit the power and elegance of the Mueller formalism we discuss the measurement of the Wolfenstein parameters by triple-scattering experiments. The initial beam is assumed to be of unit intensity and unpolarized and is thus represented by the Stokes vector

$$S_c^{(0)} = \begin{bmatrix} 1 \\ 0 \\ 0 \\ 0 \end{bmatrix}_c, \tag{3.70}$$

with the z-axis along the particle motion.

First scattering. Choosing $\phi_1 = 0$, the scattered beam has the Stokes vector

$$S_{c'}^{(1)} = \tilde{Z}_{c'c}^{(1)} S_c^{(0)} = X^{(1)} \begin{bmatrix} 1 & 0 & v^{(1)} & 0 \\ 0 & W^{(1)} & 0 & -w^{(1)} \\ v^{(1)} & 0 & 1 & 0 \\ 0 & w^{(1)} & 0 & W^{(1)} \end{bmatrix}_{c'c} \begin{bmatrix} 1 \\ 0 \\ 0 \\ 0 \end{bmatrix}_c = X^{(1)} \begin{bmatrix} 1 \\ 0 \\ v^{(1)} \\ 0 \end{bmatrix}_{c'}, \tag{3.71}$$

so that the final polarization vector is perpendicular to the scattering plane.

Second scattering. For arbitrary ϕ_2, the twice-scattered beam is described by

$$S_{c''}^{(2)} = \tilde{Z}_{c''c'}^{(2)} S_{c'}^{(1)}$$

$$= X^{(2)} \begin{bmatrix} 1 & -v^{(2)} \sin\phi_2 & v^{(2)} \cos\phi_2 & 0 \\ 0 & W^{(2)} \cos\phi_2 & W^{(2)} \sin\phi_2 & -w^{(2)} \\ v^{(2)} & -\sin\phi_2 & \cos\phi_2 & 0 \\ 0 & w^{(2)} \cos\phi_2 & w^{(2)} \sin\phi_2 & W^{(2)} \end{bmatrix}_{c''c'} X^{(1)} \begin{bmatrix} 1 \\ 0 \\ v^{(1)} \\ 0 \end{bmatrix}_{c'}$$

$$= X^{(2)} X^{(1)} \begin{bmatrix} 1 + v^{(1)} v^{(2)} \cos\phi_2 \\ v^{(1)} W^{(2)} \sin\phi_2 \\ v^{(2)} + v^{(1)} \cos\phi_2 \\ v^{(1)} w^{(2)} \sin\phi_2 \end{bmatrix}_{c''}. \tag{3.72}$$

Thus the intensity after double-scattering is, as before,

$$I_{c''}^{(2)} = X^{(2)} X^{(1)} (1 + v^{(1)} v^{(2)} \cos\phi_2), \tag{3.73}$$

and choosing $\phi_2 = 0$ and π allows the determination of the asymmetry parameter $\varepsilon_2(\theta_1, \theta_2)$, and thus one of the vector scattering parameters if the other is known (cf. Section 3.4). For the measurement of the Wolfenstein parameter W, it is convenient to choose $\phi_2 = \frac{1}{2}\pi$, so that

$$S_{c''}^{(2)} = X^{(2)}X^{(1)}\begin{bmatrix} 1 \\ v^{(1)}W^{(2)} \\ v^{(2)} \\ v^{(1)}w^{(2)} \end{bmatrix}_{c''}. \tag{3.74}$$

Third scattering. For any angle ϕ_3 we have

$$S_{c'''}^{(3)} = X^{(3)}\begin{bmatrix} 1 & -v^{(3)}\sin\phi_3 & v^{(3)}\cos\phi_3 & 0 \\ 0 & W^{(3)}\cos\phi_3 & W^{(3)}\sin\phi_3 & -w^{(3)} \\ v^{(3)} & -\sin\phi_3 & \cos\phi_3 & 0 \\ 0 & w^{(3)}\cos\phi_3 & w^{(3)}\sin\phi_3 & W^{(3)} \end{bmatrix}_{c'''c''} \times$$

$$\times X^{(2)}X^{(1)}\begin{bmatrix} 1 \\ v^{(1)}W^{(2)} \\ v^{(2)} \\ v^{(1)}w^{(2)} \end{bmatrix}_{c''} \tag{3.75}$$

$$= X^{(3)}X^{(2)}X^{(1)}\begin{bmatrix} 1 - v^{(1)}W^{(2)}v^{(3)}\sin\phi_3 + v^{(2)}v^{(3)}\cos\phi_3 \\ v^{(1)}W^{(2)}W^{(3)}\cos\phi_3 + v^{(2)}W^{(3)}\sin\phi_3 - v^{(1)}w^{(2)}w^{(3)} \\ v^{(3)} - v^{(1)}W^{(2)}\sin\phi_3 + v^{(2)}\cos\phi_3 \\ v^{(1)}W^{(2)}w^{(3)}\cos\phi_3 + v^{(2)}w^{(3)}\sin\phi_3 + v^{(1)}w^{(2)}W^{(3)} \end{bmatrix}_{c'''}.$$

Thus the intensity after the third scattering is

$$I_{c'''}^{(3)} = X^{(3)}X^{(2)}X^{(1)}(1 - v^{(1)}W^{(2)}v^{(3)}\sin\phi_3 + v^{(2)}v^{(3)}\cos\phi_3), \tag{3.76}$$

which for $\phi_3 = \mp\frac{1}{2}\pi$ reduces to

$$I_{c'''}^{(3)} = X^{(3)}X^{(2)}X^{(1)}(1 \pm v^{(1)}W^{(2)}v^{(3)}). \tag{3.77}$$

We can define an asymmetry parameter

$$\varepsilon_3(\theta_1, \theta_2, \theta_3) = \frac{\{I_{c'''}^{(3)}(-\frac{1}{2}\pi) - I_{c'''}^{(3)}(\frac{1}{2}\pi)\}}{\{I_{c'''}^{(3)}(-\frac{1}{2}\pi) + I_{c'''}^{(3)}(\frac{1}{2}\pi)\}} = v^{(1)}W^{(2)}v^{(3)}, \tag{3.78}$$

and, provided $v^{(1)}$ and $v^{(3)}$ are known, $W^{(2)}$ can be determined.

The second Wolfenstein parameter w may be determined by a similar triple-scattering experiment provided an electric or magnetic field is used to rotate the polarization vector from a transverse direction towards a longitudinal direction between the first and second scatterings.

3.9. Polarization transfer coefficients

Many elements of the standard Mueller matrix of eqn (3.67) are closely related to quantities which are called *polarization transfer coefficients*. This is to be expected, since both the polarization transfer coefficients and the Mueller matrix elements are defined in terms of linear transformations of observables (the equivalent of the Stokes parameters). It is conventional to define the polarization transfer coefficients for $\phi = 0$, i.e. for both the y- and y'-axes perpendicular to the scattering plane (Ohlsen 1972). Then

$$\tilde{Z}_{c'c}(\phi = 0) = X \begin{bmatrix} 1 & 0 & v & 0 \\ 0 & W & 0 & -w \\ v & 0 & 1 & 0 \\ 0 & w & 0 & W \end{bmatrix}_{c'c} \equiv X \begin{bmatrix} Z_0^0 & Z_x^0 & Z_y^0 & Z_z^0 \\ Z_0^{x'} & Z_x^{x'} & Z_y^{x'} & Z_z^{x'} \\ Z_0^{y'} & Z_x^{y'} & Z_y^{y'} & Z_z^{y'} \\ Z_0^{z'} & Z_x^{z'} & Z_y^{z'} & Z_z^{z'} \end{bmatrix}_{c'c}, \quad \text{say,}$$

(3.79)

where we have re-labelled and divided all the matrix elements into four groups (1) $Z_0^0 \equiv 1$, (2) Z_x^0, Z_y^0, and Z_z^0, (3) $Z_0^{x'}$, $Z_0^{y'}$, and $Z_0^{z'}$, (4) $Z_x^{x'}$, $Z_y^{x'}$, etc. Thus for an incident beam of intensity $I^{(i)}$ and polarization components p_j, the scattered beam has intensity $I^{(f)}$ and polarization components $p_{k'}$ given by

$$I^{(f)} = I^{(i)}X \left(Z_0^0 + \sum_j p_j Z_j^0 \right) \equiv I^{(i)}X \left(1 + \sum_j p_j v_j \right),$$

(3.80)

and

$$I^{(f)}p_{k'} = I^{(i)}X \left(Z_0^{k'} + \sum_j p_j Z_j^{k'} \right).$$

(3.81)

It is usual to refer to the set (2) as (vector) analysing powers, set (3) as (vector) polarizations produced by an unpolarized incident beam, and set (4) as polarization transfer coefficients. We shall refer to the complete set of matrix elements as *Mueller coefficients*.

3.9.1. *Parity conservation*

If the scattering interaction conserves parity, many of the Mueller coefficients are identically zero. The rule is (Ohlsen 1972): 'if the number of x' and z' superscripts plus the number of x and z subscripts is *odd*, the Mueller coefficient is zero'. This follows from the observation that, for the particular choice of axes, x, x', z, and z' are linear combinations of $\mathbf{k}$ and $\mathbf{k}'$ which are 'odd' quantities under the parity operation while y and y' being along $\mathbf{k} \times \mathbf{k}'$ are 'even' quantities. Hence

$$Z_z^0 = Z_x^0 = Z_0^{z'} = Z_0^{x'} = Z_y^{z'} = Z_y^{x'} = Z_z^{y'} = Z_x^{y'} = 0,$$

(3.82)

as already deduced.

3.9.2. *Odd-even character of Mueller coefficients*

Rotational invariance requires that 'the Mueller coefficients are odd or even functions of the scattering angle θ according to the number of x' and y' superscripts plus the number of x and y subscripts is odd or even'. This follows (see Ohlsen 1972) from invariance under a rotation of 180° about the z-axis. Thus Z_y^0 and $Z_0^{y'}$ are odd functions of θ while $Z_x^{x'}$, $Z_z^{x'}$, $Z_y^{y'}$, $Z_x^{z'}$, and $Z_z^{z'}$ are even functions of θ.

3.9.3. *Relations between Mueller coefficients*

For the elastic scattering of spin-$\frac{1}{2}$ particles by a spinless target we have the following relations (from eqn (3.79)) if parity is conserved

$$Z_y^0 = Z_0^{y'} = v, \tag{3.83a}$$

$$Z_y^{y'} = 1, \tag{3.83b}$$

$$Z_z^{z'} = Z_x^{x'} = W, \tag{3.83c}$$

and

$$Z_x^{z'} = -Z_z^{x'} = w. \tag{3.83d}$$

3.9.4. *Formulae for Mueller coefficients*

Using the relations $I^{(\mathrm{f})} = \mathrm{tr}\,\rho^{(\mathrm{f})} = \mathrm{tr}(\tilde{M}\rho^{(i)}\tilde{M}^\dagger)$ and $\rho^{(i)} = \frac{1}{2}I^{(i)}(1 + \sum_j p_j \hat{\sigma}_j)$, where $\tilde{M}$ is the standard scattering matrix of eqn (3.53) with $\phi_q = 0$ and for convenience the subscripts c and c' defining the standard axes have been omitted, we have

$$I^{(\mathrm{f})} = \tfrac{1}{2}I^{(i)}\left\{\mathrm{tr}(\tilde{M}\tilde{M}^\dagger) + \sum_j p_j\,\mathrm{tr}(\tilde{M}\hat{\sigma}_j\tilde{M}^\dagger)\right\}$$

$$= I^{(i)}X\left\{1 + \sum_j p_j\frac{\mathrm{tr}(\tilde{M}\hat{\sigma}_j\tilde{M}^\dagger)}{\mathrm{tr}(\tilde{M}\tilde{M}^\dagger)}\right\}. \tag{3.84}$$

By comparison with eqn (3.80),

$$Z_j^0 = \mathrm{tr}(\tilde{M}\hat{\sigma}_j\tilde{M}^\dagger)/\mathrm{tr}(\tilde{M}\tilde{M}^\dagger). \tag{3.85}$$

Similarly, using eqn (3.40), we have

$$I^{(\mathrm{f})}p_{k'} = \mathrm{tr}(\rho^{(\mathrm{f})}\hat{\sigma}_{k'}) = \mathrm{tr}(\tilde{M}\rho^{(i)}\tilde{M}^\dagger\hat{\sigma}_{k'})$$

$$= \tfrac{1}{2}I^{(i)}\left\{\mathrm{tr}(\tilde{M}\tilde{M}^\dagger\hat{\sigma}_{k'}) + \sum_j p_j\,\mathrm{tr}(\tilde{M}\hat{\sigma}_j\tilde{M}^\dagger\hat{\sigma}_{k'})\right\}$$

$$= I^{(i)}X\left\{\mathrm{tr}(\tilde{M}\tilde{M}^\dagger\hat{\sigma}_{k'}) + \sum_j p_j\,\mathrm{tr}(\tilde{M}\hat{\sigma}_j\tilde{M}^\dagger\hat{\sigma}_{k'})\right\}/\mathrm{tr}(\tilde{M}\tilde{M}^\dagger). \tag{3.86}$$

Thus, by comparison with eqn (3.81),

$$Z_0^{k'} = \text{tr}(\tilde{M}\tilde{M}^\dagger\hat{\sigma}_{k'})/\text{tr}(\tilde{M}\tilde{M}^\dagger) \tag{3.87}$$

and

$$Z_j^{k'} = \text{tr}(\tilde{M}\hat{\sigma}_j\tilde{M}^\dagger\hat{\sigma}_{k'})/\text{tr}(\tilde{M}\tilde{M}^\dagger). \tag{3.88}$$

4

SCATTERING OF SPIN-1 PARTICLES

IN this chapter we discuss, in the non-relativistic limit, the elastic scattering of spin-1 particles. To be specific we shall consider deuterons, but our treatment applies generally to particles of spin-1 which have non-zero mass. As discussed in Chapter 2, the massless photons constitute a special case.

4.1. Polarized deuterons

Deuterons have spin 1 and consequently are associated with an intrinsic angular momentum operator $\hat{S}$ which satisfies the usual commutation relations (2.23) with

$$\hat{S}^2 = \hat{S}_x^2 + \hat{S}_y^2 + \hat{S}_z^2 = 2\hbar^2 \hat{1}_3. \tag{4.1}$$

Thus deuterons can exist in three basic spin states corresponding to the eigenvalues $m_s = 0, \pm\hbar$ of one of the components (usually $\hat{S}_z$) of the operator $\hat{S}$. A polarized deuteron beam may be represented by a three-component spin wave function

$$\chi_c = \begin{bmatrix} a_1 \\ a_2 \\ a_3 \end{bmatrix}_c = \sum_n a_n \phi_c^{(n)}, \tag{4.2}$$

where

$$\phi_c^{(1)} = \begin{bmatrix} 1 \\ 0 \\ 0 \end{bmatrix}_c, \qquad \phi_c^{(2)} = \begin{bmatrix} 0 \\ 1 \\ 0 \end{bmatrix}_c, \qquad \phi_c^{(3)} = \begin{bmatrix} 0 \\ 0 \\ 1 \end{bmatrix}_c \tag{4.3}$$

are a diagonalized basis (see Section 2.4.1) and the subscript c denotes a system of reference axes. The spin wave function χ_c is the analogue of the Jones vector J_c of eqn (1.13) and a generalization of the spin-$\frac{1}{2}$ wave function of eqn (3.1). For a completely polarized beam and the basis $\phi_c^{(n)}$ we can form a density matrix from the wave function, i.e.

$$\rho_c = \chi_c \chi_c^\dagger = \begin{bmatrix} |a_1|^2 & a_1 a_2^* & a_1 a_3^* \\ a_1^* a_2 & |a_2|^2 & a_2 a_3^* \\ a_1^* a_3 & a_2^* a_3 & |a_3|^2 \end{bmatrix}_c, \tag{4.4}$$

which contains *five* independent quantities corresponding to the three real and the three imaginary parts of the amplitudes a_n -1 for the over-all indeterminate phase of the wave function χ_c (alternatively, the three moduli $|a_n|$ plus the two relative phases of the amplitudes a_2 and a_3 to the amplitude a_1).

In the general case of a partially polarized deuteron beam we have a 3×3 Hermitian density matrix

$$\rho_c = \begin{bmatrix} \rho_{11} & \rho_{12} & \rho_{13} \\ \rho_{21} & \rho_{22} & \rho_{23} \\ \rho_{31} & \rho_{32} & \rho_{33} \end{bmatrix}_c = \begin{bmatrix} \rho_{11} & \rho_{12} & \rho_{13} \\ \rho_{12}^* & \rho_{22} & \rho_{23} \\ \rho_{13}^* & \rho_{23}^* & \rho_{33} \end{bmatrix}_c , \tag{4.5}$$

which has *nine* independent quantities: the three real diagonal elements plus the real and imaginary parts of ρ_{12}, ρ_{13}, and ρ_{23}. The diagonal elements again give the relative probabilities of a deuteron being found in the respective basis states, and the trace is a measure of the beam intensity.

For a representation of the spin operators, it is convenient to choose $\hat{S}_z$ in diagonal form with the $\phi_c^{(n)}$ as the eigenstates χ_{1m_s}, i.e.

$$\hat{S}_z = \hbar \begin{bmatrix} 1 & 0 & 0 \\ 0 & 0 & 0 \\ 0 & 0 & -1 \end{bmatrix} \tag{4.6}$$

and

$$\phi_c^{(1)} = \chi_{11}, \qquad \phi_c^{(2)} = \chi_{10}, \qquad \phi_c^{(3)} = \chi_{1\,-1}. \tag{4.7}$$

Using the commutation relations and eqn (4.6) we obtain $\hat{S}_x$ and $\hat{S}_y$ of the form

$$\hat{S}_x = \hbar \begin{bmatrix} 0 & b & 0 \\ d & 0 & f \\ 0 & h & 0 \end{bmatrix}, \qquad \hat{S}_y = -i\hbar \begin{bmatrix} 0 & b & 0 \\ -d & 0 & f \\ 0 & -h & 0 \end{bmatrix} \tag{4.8}$$

with the constraint $bd = fh = \frac{1}{2}$. Writing (cf. eqn (2.47))

$$\hat{\mathbf{S}} = \hbar\hat{\boldsymbol{\sigma}} \tag{4.9}$$

and making the usual choice $b = d = f = h = 1/\sqrt{2}$, we obtain

$$\hat{\sigma}_x = \frac{1}{\sqrt{2}} \begin{bmatrix} 0 & 1 & 0 \\ 1 & 0 & 1 \\ 0 & 1 & 0 \end{bmatrix}, \qquad \hat{\sigma}_y = \frac{1}{\sqrt{2}} \begin{bmatrix} 0 & -i & 0 \\ i & 0 & -i \\ 0 & i & 0 \end{bmatrix},$$

$$\hat{\sigma}_z = \begin{bmatrix} 1 & 0 & 0 \\ 0 & 0 & 0 \\ 0 & 0 & -1 \end{bmatrix}. \tag{4.10}$$

These operators are the analogues of the Pauli spin matrices for spin-$\frac{1}{2}$ particles and are very similar to the matrices of eqns (2.36) and (2.38) which correspond to a different choice of b and d. Clearly, the matrix operators $\hat{1}_3$, $\hat{\sigma}_x$, $\hat{\sigma}_y$, $\hat{\sigma}_z$ do not give a complete representation of the general 3×3 density matrix for spin-1 particles. It is conventional to use either the following set of spherical tensor operators (Lakin 1955):

$$\hat{\tau}_{00} = \hat{1}_3, \tag{4.11a}$$

$$\hat{\tau}_{10} = \sqrt{(\tfrac{3}{2})}\hat{\sigma}_z, \tag{4.11b}$$

$$\hat{\tau}_{11} = -\tfrac{1}{2}\sqrt{(3)}(\hat{\sigma}_x + i\hat{\sigma}_y), \tag{4.11c}$$

$$\hat{\tau}_{20} = \tfrac{1}{2}\sqrt{(2)}(3\hat{\sigma}_z^2 - 2\hat{1}_3), \tag{4.11d}$$

$$\hat{\tau}_{21} = -\tfrac{1}{2}\sqrt{(3)}\{(\hat{\sigma}_x + i\hat{\sigma}_y)\hat{\sigma}_z + \hat{\sigma}_z(\hat{\sigma}_x + i\hat{\sigma}_y)\}, \tag{4.11e}$$

$$\hat{\tau}_{22} = \tfrac{1}{2}\sqrt{(3)}(\hat{\sigma}_x + i\hat{\sigma}_y)^2 ; \tag{4.11f}$$

and $\hat{\tau}_{1-1}, \hat{\tau}_{2-1}$, and $\hat{\tau}_{2-2}$ defined by

$$\hat{\tau}_{K-Q} = (-1)^Q \hat{\tau}_{KQ}^\dagger, \tag{4.12}$$

or the set of Cartesian tensor operators (Goldfarb 1958)

$$\hat{\mathcal{P}}_i = \hat{\sigma}_i, \quad i = x, y, \text{ or } z, \tag{4.13a}$$

$$\hat{\mathcal{P}}_{ij} = \tfrac{3}{2}(\hat{\sigma}_i\hat{\sigma}_j + \hat{\sigma}_j\hat{\sigma}_i) - 2\delta_{ij}\hat{1}_3, \quad i,j = x, y, \text{ or } z, \tag{4.13b}$$

where $\delta_{ij} = 0$ if $i \neq j$ and $\delta_{ij} = 1$ if $i = j$.

The use of Cartesian tensor operators has some advantages as stressed by Ohlsen (1972), but for our purposes we find it more convenient to employ the spherical tensor operators which are irreducible and obey the well-known transformation rule for rotations from axes c to axes c':

$$(\hat{\tau}_{KQ'})_{c'} = \sum_Q (\hat{\tau}_{KQ})_c D_{QQ'}^{(K)}(\alpha, \beta, \gamma), \tag{4.14}$$

where $D_{QQ'}^{(K)}$ are the rotation matrix elements and α, β, γ are the Euler angles as defined by Brink and Satchler (1968). The matrix elements are given by

$$D_{QQ'}^{(K)} = e^{-i\alpha Q} d_{QQ'}^{(K)}(\beta) e^{-i\gamma Q'}, \tag{4.15}$$

where

$$d_{QQ'}^{(K)}(\beta) = \sum_n (-1)^n \frac{\{(K+Q)!(K-Q)!(K+Q')!(K-Q')!\}^{\frac{1}{2}}}{(K+Q-n)!(K-Q'-n)!n!(n+Q'-Q)!} \times$$
$$\times (\cos \tfrac{1}{2}\beta)^{2K+Q-Q'-2n}(\sin \tfrac{1}{2}\beta)^{2n+Q'-Q}. \tag{4.16}$$

The expectation values of the operators $\hat{\tau}_{KQ}$ are defined in terms of the density matrix (omitting the reference axes)

$$t_{KQ} \equiv \langle \hat{\tau}_{KQ} \rangle = \text{tr}(\rho\hat{\tau}_{KQ})/\text{tr } \rho, \tag{4.17}$$

and under a rotation of axes from eqn (4.14) we have

$$(t_{KQ'})_{c'} = \sum_Q (t_{KQ})_c D_{QQ'}^{(K)}(\alpha, \beta, \gamma). \tag{4.18}$$

Using the nine equations (4.17) we can express the elements of the density matrix in terms of the expectation values t_{KQ}; we find

$$\rho_c = \tfrac{1}{3}I \times$$

$$\times \begin{bmatrix} 1 + \sqrt{(\tfrac{3}{2})}t_{10} + \dfrac{1}{\sqrt{(2)}}t_{20} & \sqrt{(\tfrac{3}{2})}t_{1\,-1} + \sqrt{(\tfrac{3}{2})}t_{2\,-1} & \sqrt{(3)}t_{2\,-2} \\[2mm] -\sqrt{(\tfrac{3}{2})}t_{11} - \sqrt{(\tfrac{3}{2})}t_{21} & 1 - \sqrt{(2)}t_{20} & \sqrt{(\tfrac{3}{2})}t_{1\,-1} - \sqrt{(\tfrac{3}{2})}t_{2\,-1} \\[2mm] \sqrt{(3)}t_{22} & -\sqrt{(\tfrac{3}{2})}t_{11} + \sqrt{(\tfrac{3}{2})}t_{21} & 1 - \sqrt{(\tfrac{3}{2})}t_{10} + \dfrac{1}{\sqrt{(2)}}t_{20} \end{bmatrix}_c ,$$

$$\tag{4.19}$$

which may be written as

$$\rho_c = \tfrac{1}{3}I \left(\sum_{KQ} t_{KQ} \hat{t}_{KQ}^{\dagger} \right)_c . \tag{4.20}$$

This expression gives the polarization state of the deuteron beam and is the analogue of eqn (3.32) for spin-$\tfrac{1}{2}$ particles. To specify the polarization completely we must know the values of all nine t_{KQ}. The t_{1Q} and t_{2Q} expectation values are known as *vector* and *tensor* (rank 2) polarizations, respectively.

It is worthwhile noting several fundamental differences between the polarizations of spin-1 and spin-$\tfrac{1}{2}$ particles. For spin-$\tfrac{1}{2}$ particles it is possible to consider an arbitrary beam as an incoherent mixture of a polarized beam (three parameters) and an unpolarized beam (one parameter). However, for an arbitrary deuteron beam we need nine independent quantities to specify the density matrix, and consequently we cannot generally express such a beam as a mixture of an unpolarized beam (one parameter) and a polarized beam (five parameters). For spin-$\tfrac{1}{2}$ particles, it is also possible to represent (although not uniquely) the density matrix as a sum (as in eqn (2.41)) of two completely polarized beams, e.g.

$$\rho_c = \begin{bmatrix} |a_1|^2 & a_1 a_2^* \\ a_1^* a_2 & |a_2|^2 \end{bmatrix}_c + \begin{bmatrix} |b_1|^2 & 0 \\ 0 & 0 \end{bmatrix}_c . \tag{4.21}$$

Analogously, the density matrix of an arbitrary deuteron beam may be expressed as the sum of three completely polarized beams, namely,

$$\rho_c = \begin{bmatrix} |a_1|^2 & a_1 a_2^* & a_1 a_3^* \\ a_1^* a_2 & |a_2|^2 & a_2 a_3^* \\ a_1^* a_3 & a_2^* a_3 & |a_3|^2 \end{bmatrix}_c + \begin{bmatrix} |b_1|^2 & 0 & b_1 b_3^* \\ 0 & 0 & 0 \\ b_1^* b_3 & 0 & |b_3|^2 \end{bmatrix}_c + \begin{bmatrix} 0 & 0 & 0 \\ 0 & |d_2|^2 & 0 \\ 0 & 0 & 0 \end{bmatrix}_c . \tag{4.22}$$

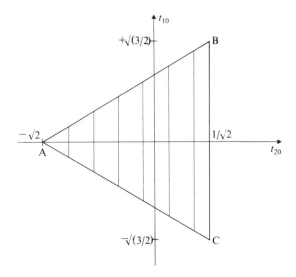

FIG. 4.1. The allowed values of t_{10}, t_{20} correspond to points inside and on the triangle ABC. Points A, B, C are states of complete polarization ($D = 1$).

Five parameters are associated with the first matrix, three with the second, and one with the third, making a total of nine parameters.

For spin-1 particles it is possible to have a completely polarized beam which has no preferred 'spin' direction, i.e. $t_{1\pm1} = t_{10} = 0$ and is a pure tensor polarized beam. For example, a beam described by the wave function χ_{10} has density matrix

$$\rho_{c} = \begin{bmatrix} 0 & 0 & 0 \\ 0 & 1 & 0 \\ 0 & 0 & 0 \end{bmatrix}_{c}, \qquad (4.23)$$

and comparison of eqns (4.23) and (4.19) shows that all the t_{1Q}, t_{2Q} are zero except $t_{20} = -\sqrt{2}$. Conversely, a pure vector polarized beam (i.e. all $t_{2Q} = 0$) cannot be represented by a wave function and hence is only partially polarized. Finally, an equal mixture of the opposite spin states χ_{11} and χ_{1-1} leads to a density matrix which is not equal to the density matrix for an unpolarized beam, i.e.

$$\rho_{c} = \begin{bmatrix} \frac{1}{2} & 0 & 0 \\ 0 & 0 & 0 \\ 0 & 0 & \frac{1}{2} \end{bmatrix}_{c} \neq \rho_{unpol} = \begin{bmatrix} \frac{1}{3} & 0 & 0 \\ 0 & \frac{1}{3} & 0 \\ 0 & 0 & \frac{1}{3} \end{bmatrix}_{c}. \qquad (4.24)$$

4.1.1. *Degree of polarization*

A completely polarized beam of spin-$\frac{1}{2}$ particles may be represented by a wave function (i.e. a pure state), and has polarization $\mathbf{p}$ of unit magnitude; partially polarized beams have 'degree of polarization' $|\mathbf{p}| < 1$. For deuteron beams we have similar statements: a completely polarized beam may be described by a wave function and has vector and tensor polarizations such that the *degree of polarization*

$$D = \left(\tfrac{1}{2} \sum_{K=1,2} \sum_Q |t_{KQ}|^2\right)^{\frac{1}{2}} = 1; \tag{4.25}$$

partially polarized beams have $D < 1$.

4.1.2. *Limitations on magnitudes of t_{KQ}*

(a) t_{10}, t_{20}. The diagonal elements of the density matrix which may be written in terms of t_{10} and t_{20} must be positive definite and this restricts the possible values of t_{10} and t_{20}. For a beam of unit intensity we require

$$\left.\begin{aligned}
\rho_{11} &= \tfrac{1}{3}\left(1 + \sqrt{(\tfrac{3}{2})}t_{10} + \frac{1}{\sqrt{2}}t_{20}\right) \geqslant 0\\[4pt]
\rho_{22} &= \tfrac{1}{3}(1 - \sqrt{(2)}t_{20}) \qquad\qquad \geqslant 0\\[4pt]
\rho_{33} &= \tfrac{1}{3}\left(1 - \sqrt{(\tfrac{3}{2})}t_{10} + \frac{1}{\sqrt{2}}t_{20}\right) \geqslant 0
\end{aligned}\right\} \tag{4.26}$$

These three conditions must be satisfied simultaneously, and we find

$$-\sqrt{\tfrac{3}{2}} \leqslant t_{10} \leqslant \sqrt{\tfrac{3}{2}} \tag{4.27a}$$

and

$$-\sqrt{2} \leqslant t_{20} \leqslant \frac{1}{\sqrt{2}}. \tag{4.27b}$$

Both t_{10} and t_{20} do not take their extremum values simultaneously; the allowed combinations are shown in Fig. 4.1.

(b) $t_{1\pm1}, t_{2\pm1}, t_{2\pm2}$. The quantities t_{11} and t_{21} are related to ρ_{21} and ρ_{32} in the density matrix of eqn (4.5) which when expanded in the form (4.22) gives

$$-\frac{1}{\sqrt{6}}(t_{11} + t_{21}) = a_1^* a_2 = r_1 r_2 \, e^{i(\delta_2 - \delta_1)} \tag{4.28a}$$

and

$$-\frac{1}{\sqrt{6}}(t_{11} - t_{21}) = a_2^* a_3 = r_2 r_3 \, e^{i(\delta_3 - \delta_2)}, \tag{4.28b}$$

where the complex coefficients a_n have been written in the form $r_n \, e^{i\delta_n}$. Adding eqns (4.28a) and (4.28b) gives

$$t_{11} = -\tfrac{1}{2}\sqrt{(6)}(r_1 \, e^{i(\delta_2 - \delta_1)} + r_3 \, e^{i(\delta_3 - \delta_2)})r_2. \qquad (4.29)$$

The maximum value of $|t_{11}|$ occurs when both phase factors are equal so that

$$|t_{11}|_{\text{max}} = \tfrac{1}{2}\sqrt{(6)}\{(r_1 + r_3)r_2\}_{\text{max}}, \qquad (4.30)$$

where for a beam of unit intensity $r_1^2 + r_2^2 + r_3^2 \leqslant 1$. The expression $(r_1 + r_3)r_2$ has a maximum value of $\tfrac{1}{2}\sqrt{2}$ when $r_2 = \tfrac{1}{2}\sqrt{2}$, $r_1 = r_3 = \tfrac{1}{2}$. Thus

$$0 \leqslant |t_{11}| \leqslant \tfrac{1}{2}\sqrt{3}. \qquad (4.31)$$

In a similar way, we may show that $|t_{1-1}|$, $|t_{2\pm 1}|$ and $|t_{2\pm 2}|$ satisfy the same relation (4.31).

4.1.3. Case of photons

The transverse nature of light arising from the non-existence of the state $\phi_c^{(2)}$ of eqn (4.3) implies (from eqn (4.19)) that for photons we have

$$t_{20} = \frac{1}{\sqrt{2}} \quad \text{and} \quad t_{1\pm 1} = t_{2\pm 1} = 0,$$

so that the density matrix is

$$\rho_c^{(\gamma)} = \frac{I}{3} \begin{bmatrix} \tfrac{3}{2} + \sqrt{(\tfrac{3}{2})}t_{10} & 0 & \sqrt{(3)}t_{2-2} \\ 0 & 0 & 0 \\ \sqrt{(3)}t_{22} & 0 & \tfrac{3}{2} - \sqrt{(\tfrac{3}{2})}t_{10} \end{bmatrix}_c. \qquad (4.32)$$

Comparing the non-vanishing elements of $\rho_c^{(\gamma)}$ with the density matrix of eqn (1.51) gives for the Stokes parameters

$$P_1 = \sqrt{(\tfrac{2}{3})}It_{10}, \quad P_2 \pm iP_3 = \frac{2}{\sqrt{(3)}}It_{2\pm 2}. \qquad (4.33)$$

For photons, the degree of polarization $D^{(\gamma)}$ (defined by eqn (4.25)) satisfies $\tfrac{1}{2} \leqslant D^{(\gamma)} \leqslant 1$, so that so-called 'unpolarized' light is, strictly speaking, partially tensor polarized ($t_{20} = 1/\sqrt{2}$) with $D^{(\gamma)} = \tfrac{1}{2}$.

4.2. Elastic scattering of deuterons by a spinless target

As for spin-$\tfrac{1}{2}$ particles the density matrix for an elastically scattered deuteron beam is given in terms of the density matrix describing the incident beam $\rho_c^{(i)}$ and the scattering matrix $\hat{M}_{cc}$ describing the interaction for c.m. coordinate system c, i.e.

$$\rho_c^{(f)} = \hat{M}_{cc}\rho_c^{(i)}\hat{M}_{cc}^\dagger, \qquad (4.34)$$

where all the quantities are 3×3 matrices. The general form of the scattering matrix assuming rotational invariance (i.e. each term is either scalar or pseudoscalar), parity conservation, and reciprocity (see Section 3.1) may be written as

$$\hat{M}_{cc} = a\hat{1}_3 + b(\mathbf{N} \cdot \hat{\boldsymbol{\sigma}}) + c(\mathbf{M} \cdot \hat{\boldsymbol{\sigma}})^2 + d(\mathbf{K} \cdot \hat{\boldsymbol{\sigma}})^2, \tag{4.35}$$

where a, b, c, d are complex quantities which are functions of energy and angular variables analogous to g, h of eqn (3.7) for spin-$\frac{1}{2}$ particles, and $\mathbf{M}, \mathbf{N}, \mathbf{K}$ are the unit vectors defined by eqn (3.5). For the z-axis along the incident beam direction $\mathbf{k}$ and arbitrary scattering direction (θ, ϕ) in the c.m. system, the scattering matrix is

$$\hat{M}_{cc} = \begin{bmatrix} A & B\,e^{-i\phi} & C\,e^{-2i\phi} \\ D\,e^{i\phi} & E & -D\,e^{-i\phi} \\ C\,e^{2i\phi} & -B\,e^{i\phi} & A \end{bmatrix}_{cc}, \tag{4.36}$$

where

$$A = \tfrac{1}{4}\{4a + (3 - \cos\theta)c + (3 + \cos\theta)d\}, \tag{4.37a}$$

$$B = \frac{1}{2\sqrt{2}}\{-2ib - (c - d)\sin\theta\}, \tag{4.37b}$$

$$C = \tfrac{1}{4}\{(1 + \cos\theta)c + (1 - \cos\theta)d\}, \tag{4.37c}$$

$$D = B + \sqrt{(2)}ib, \tag{4.37d}$$

and

$$E = \tfrac{1}{2}\{2a + (1 + \cos\theta)c + (1 - \cos\theta)d\} = (A - C) - \sqrt{(2)}(B + D)\cot\theta. \tag{4.38}$$

Eqn (4.36) is analogous to eqn (3.26) for spin-$\frac{1}{2}$ particles. The four amplitudes A, B, C, D are independent, so that, neglecting an over-all phase factor, there are in principle *seven* independent combinations of these amplitudes which are observable. Thus a minimum number of seven measurements (at each energy and for each scattering angle) are essential to determine the matrix $\hat{M}_{cc}$ completely, and because of the bilinear dependence of the observables on the amplitudes a larger number of experiments may be required in practice to eliminate spurious solutions.

4.2.1. Deuteron optical potential

As for nucleons, the elastic scattering of deuterons by complex nuclei is often described in terms of an optical potential (Hodgson 1966). In this model the deuteron–nucleus interaction is spin-dependent and (neglecting any target spin interactions) is analogous to that of eqn (3.9) for nucleons, except (Satchler 1960) that it contains three possible second-rank tensor terms in addition to the usual central and spin–orbit potentials. Under parity conservation and the requirement of reciprocity, the tensor potentials are of the form

$$T_R = \{(\hat{\boldsymbol{\sigma}} \cdot \mathbf{r})^2 r^{-2} - \tfrac{2}{3}\} V_R(r), \tag{4.39}$$

$$T_L = \{(\hat{\boldsymbol{\sigma}} \cdot \hat{\boldsymbol{\lambda}})^2 + \tfrac{1}{2}(\hat{\boldsymbol{\sigma}} \cdot \hat{\boldsymbol{\lambda}}) - \tfrac{2}{3}\hat{\boldsymbol{\lambda}}^2\} V_L(r), \tag{4.40}$$

$$T_P = \{(\hat{\boldsymbol{\sigma}} \cdot \hat{\mathbf{P}})^2 - \tfrac{2}{3}\hat{\mathbf{P}}^2\} V_P(r) + V_P(r)\{(\hat{\boldsymbol{\sigma}} \cdot \hat{\mathbf{P}})^2 - \tfrac{2}{3}\hat{\mathbf{P}}^2\}, \tag{4.41}$$

where $\hat{\boldsymbol{\sigma}} = \hat{\mathbf{S}}/\hbar$, $\hat{\boldsymbol{\lambda}} = \hat{\mathbf{L}}/\hbar$, $\hat{\mathbf{P}} = -i\boldsymbol{\nabla}$ and V_R, V_L, V_P are radial form factors which may have imaginary parts. A complication which arises with the tensors T_R and T_P is that both couple certain partial waves differing by two units of orbital angular momentum.

4.2.2. *Partial wave analysis for deuteron optical potential with tensor terms*

As an example of the relationship between a theoretical model and the scattering matrix we consider a deuteron–nucleus potential of the form

$$V(r) = V_C(r) + S(r)\hat{\boldsymbol{\sigma}} \cdot \hat{\boldsymbol{\lambda}} + T_R + T_L + T_P. \tag{4.42}$$

The general solution of the Schrödinger equation for this interaction for a completely polarized incident beam with spin component m_s may be written as a partial wave expansion (analogous to eqn (3.16))

$$\psi_{m_s} = \sum_{jll'm_j} A_{jlm_jm_s} R_{jll'}(r) F_{jl'm_j}, \tag{4.43}$$

where

$$F_{jl'm_j} = \sum_{m'_s} C(l'1j, m_j - m'_s m'_s m_j) Y_{l'm_j - m'_s}(\theta, \phi) \chi_{1m'_s} \tag{4.44}$$

are spin–angle functions and $R_{jll'}(r)$ are radial wave functions for incoming and outgoing orbital angular momenta, l and l', respectively. For large values of r

$$\psi_{m_s} \simeq \left[\left(1 + \frac{\eta^2}{i(kr - \mathbf{k} \cdot \mathbf{r})}\right) \exp\{i\mathbf{k} \cdot \mathbf{r} + i\eta \ln(kr - \mathbf{k} \cdot \mathbf{r})\} \right] \chi_{1m_s} +$$

$$+ \sum_{m'_s} M_{m'_s m_s}(\theta, \phi) r^{-1} \exp(ikr - i\eta \ln 2kr + 2i\sigma_0) \chi_{1m'_s} \tag{4.45}$$

where $M_{m'_s m_s}$ are the elements of the scattering matrix. The $A_{jlm_jm_s}$ of eqn (4.43) which satisfy the boundary condition (4.45) are given by

$$A_{jlm_jm_s} = 4\pi i^l Y^*_{lm_j - m_s}(\theta_k, \phi_k) C(l1j, m_j - m_s m_s m_j), \tag{4.46}$$

where θ_k, ϕ_k refer to the vector $\mathbf{k}$. If the z-axis is chosen along the incident direction, m_j only takes the value m_s, and if the y-axis is taken parallel to

$\mathbf{k} \times \mathbf{k}'$ where $\mathbf{k}'$ is the final momentum, so that $\phi = 0$, the scattering matrix has elements

$$M_{m_s'm_s}(\theta) = \delta_{m_s m_s'} f_C(\theta) + \frac{1}{2ik} \sum_l \{4\pi(2l+1)\}^{\frac{1}{2}} \times$$

$$\times \left\{ \sum_{jl'} \Delta_{jll'} C(l1j, 0m_s m_s) C(l'1j, m_s - m_s'm_s'm_s) Y_{l'm_s - m_s'}(\theta, 0) i^{l-l'} \right.$$

$$\left. - \delta_{m_s m_s'} \exp(2i\omega_l) Y_{l0}(\theta, 0) \right\}, \tag{4.47}$$

where

$$\omega_l = \sum_{l''=1}^{l} \tan^{-1}(\eta/l'') \tag{4.48}$$

$$f_C = \frac{-\eta}{2k} \operatorname{cosec}^2(\tfrac{1}{2}\theta) \exp\{-2i\eta \ln \sin(\tfrac{1}{2}\theta)\}, \tag{4.49}$$

and $\Delta_{jll'}$ is given by the asymptotic relations

$$R_{jll'} \underset{r \to \infty}{\simeq} (kr)^{-1} \{\exp(-2i\omega_{l'})\Delta_{jll'} A_{l'} + \delta_{ll'} B_l\} \exp(i\omega_{l'}), \tag{4.50}$$

with $A_l = \frac{1}{2}(F_l - iG_l)$ and $B_l = \frac{1}{2}(F_l + iG_l)$, where F_l, G_l are the standard Coulomb wave functions.

The elements of the scattering matrix satisfy the relationship (from eqn (4.47))

$$M_{-m_s-m_s'}(\theta) = (-1)^{m_s - m_s'} M_{m_s m_s'}(\theta), \tag{4.51}$$

so that $\hat{M}_{cc}$ may be written in the form

$$\hat{M}_{cc} = \begin{bmatrix} A & B & C \\ D & E & -D \\ C & -B & A \end{bmatrix}_{cc}, \tag{4.52}$$

which is of the form of eqn (4.36) for the axes c defined above. Using eqn (4.47) one can verify that the amplitude E may be expressed in terms of the amplitudes A, B, C, D according to eqn (4.38). These amplitudes are functions of $\Delta_{jll'}$ which are determined by solving the coupled radial equations

$$\left\{ \frac{d^2}{dr^2} + k^2 - V_{jl'} - l'(l'+1)r^{-2} \right\} (rR_{jll'}) = (\delta_{jl+1} + \delta_{jl-l}) V_j^t (rR_{jll''}), \tag{4.53}$$

where

$$V_{jl} = 2\mu\hbar^{-2} \{ V_C(r) + S(r) K_{jl}(S) + V_R(r) K_{jl}(R) + V_L(r) K_{jl}(L) + V_P(r) K_{jl}(P) +$$

$$+ K_{jl}(P) V_P(r) \}, \tag{4.54}$$

$$V_j^t = 2\mu\hbar^{-2} \{ V_R(r) K_j^t(R) + V_P(r) K_j^t(P) + K_j^t(P) V_P(r) \}, \tag{4.55}$$

and $l'' = 2j - l'$. The quantities K_{jl} have the values in Table 4.1 for $j = l+1$, l, and $l-1$, respectively, and

$$K_j^t(R) = \{j(j+1)\}^{\frac{1}{2}}/(2j+1),$$ (4.56)

$$K_j^t(P) = -K_j^t(R)\left\{\frac{d^2}{dr^2} - \frac{(2j-1)}{r}\frac{d}{dr} + \frac{(j^2-1)}{r^2}\right\},$$ (4.57)

TABLE 4.1

Eigenvalues K_{jl}

j	$K_{jl}(S)$	$3K_{jl}(R)$	$6K_{jl}(L)$	$\frac{3}{2}(2l+1)^{\frac{1}{2}}K_{jl}(P)$
$l+1$	l	$-l(2l+3)^{-1}$	$l(2l-1)$	$-l(2l+3)^{-1}\nabla^2$
l	-1	$+1$	$-(2l+3)(2l-1)$	$-\nabla^2$
$l-1$	$-(l+1)$	$-(l+1)(2l-1)^{-1}$	$(2l+3)(l+1)$	$-(l+1)(2l-1)^{-1}\nabla^2$

and

$$\nabla^2 = \frac{d^2}{dr^2} + \frac{2}{r}\frac{d}{dr} - \frac{l(l+1)}{r^2}.$$ (4.58)

4.2.3. Scattering of an unpolarized beam

For an unpolarized beam of unit intensity, the density matrix is

$$\rho_{\text{unpol}} = \begin{bmatrix} \frac{1}{3} & 0 & 0 \\ 0 & \frac{1}{3} & 0 \\ 0 & 0 & \frac{1}{3} \end{bmatrix}.$$ (4.59)

The density matrix after the first scattering is

$$\rho_c^{(1)} = \hat{M}_{cc}\rho_{\text{unpol}}\hat{M}_{cc}^\dagger = \frac{1}{3}\hat{M}_{cc}\hat{M}_{cc}^\dagger$$

$$= \frac{1}{3}\begin{bmatrix} A & Be^{-i\phi} & Ce^{-2i\phi} \\ De^{i\phi} & E & -De^{-i\phi} \\ Ce^{2i\phi} & -Be^{i\phi} & A \end{bmatrix}_{cc}\begin{bmatrix} A^* & D^*e^{-i\phi} & C^*e^{-2i\phi} \\ B^*e^{i\phi} & E^* & -B^*e^{-i\phi} \\ C^*e^{2i\phi} & -D^*e^{i\phi} & A^* \end{bmatrix}_{cc}$$

$$= \frac{1}{3}\begin{bmatrix} (|A|^2+|B|^2+|C|^2) & (AD^*+BE^*-CD^*)e^{-i\phi} \\ (DA^*+EB^*-DC^*)e^{i\phi} & (2|D|^2+|E|^2) \\ (CA^*-|B|^2+AC^*)e^{2i\phi} & (CD^*-BE^*-AD^*)e^{i\phi} \end{bmatrix}$$

$$\begin{bmatrix} (AC^*-|B|^2+CA^*)e^{-2i\phi} \\ (DC^*-EB^*-DA^*)e^{-i\phi} \\ (|C|^2+|B|^2+|A|^2) \end{bmatrix}_c.$$ (4.60)

Hence

$$I_0 = \text{tr}\,\rho_c^{(1)} = \frac{1}{3}(2|A|^2 + 2|B|^2 + 2|C|^2 + 2|D|^2 + |E|^2),$$ (4.61)

which is the analogue of $I_0 = |g|^2 + |h|^2$ for nucleons. Comparing eqns (4.60) and (4.19) yields

$$t_{10} = 0, \tag{4.62}$$

$$I_0 t_{11} = \frac{1}{\sqrt{6}} (d_{AD} + d_{BE} - d_{CD})\, e^{i\phi}, \tag{4.63}$$

and

$$I_0 t_{20} = \tfrac{1}{3}\sqrt{(2)}(|A|^2 + |B|^2 + |C|^2 - 2|D|^2 - |E|^2), \tag{4.64}$$

$$I_0 t_{21} = -\frac{1}{\sqrt{6}} (S_{AD} + S_{BE} - S_{CD})\, e^{i\phi}, \tag{4.65}$$

$$I_0 t_{22} = \frac{1}{\sqrt{3}} (S_{AC} - |B|^2)\, e^{2i\phi}, \tag{4.66}$$

where $d_{XY} = X^*Y - Y^*X$ is pure imaginary and $S_{XY} = X^*Y + Y^*X$ is real. The polarizations t_{KQ} are usually referred to a coordinate system such that $\phi = 0$ in which case it_{11} is real and is called the *vector polarization*; t_{20}, t_{21}, and t_{22} are also real and are termed *tensor polarizations*.

4.3. Standard scattering matrix

As for spin-$\tfrac{1}{2}$ particles, it is convenient to define a standard scattering matrix $\tilde{M}_{c'c}$ for deuteron elastic scattering by a spinless target. For nucleons this matrix is given by eqn (3.53), which we can re-write in terms of the rotation matrix $D^{(\frac{1}{2})}(\phi, \theta_{\text{lab}}, 0)^\dagger$ with elements defined by eqn (4.15) for $K = \tfrac{1}{2}$. Similarly for deuterons we have a corresponding standard scattering matrix

$$\tilde{M}_{c'c} = (X/I_0)^{\frac{1}{2}} \hat{R}[\theta_{\text{lab}}] \hat{R}[\phi] \tilde{M}_{cc}(\theta, \phi)$$

$$= (X/I_0)^{\frac{1}{2}} D^{(1)}(\phi, \theta_{\text{lab}}, 0)^\dagger \hat{M}_{cc}(\theta, \phi)$$

$$= \tfrac{1}{2}(X/I_0)^{\frac{1}{2}} \begin{bmatrix} (1+\cos\theta_{\text{lab}}) & \sqrt{2}\sin\theta_{\text{lab}} & (1-\cos\theta_{\text{lab}}) \\ -\sqrt{2}\sin\theta_{\text{lab}} & 2\cos\theta_{\text{lab}} & \sqrt{2}\sin\theta_{\text{lab}} \\ (1-\cos\theta_{\text{lab}}) & -\sqrt{2}\sin\theta_{\text{lab}} & (1+\cos\theta_{\text{lab}}) \end{bmatrix} \times$$

$$\times \begin{bmatrix} e^{i\phi} & 0 & 0 \\ 0 & 1 & 0 \\ 0 & 0 & e^{-i\phi} \end{bmatrix} \hat{M}_{cc}$$

$$= (X/I_0)^{\frac{1}{2}} \begin{bmatrix} \tilde{A}\,e^{i\phi} & \tilde{B} & \tilde{C}\,e^{-i\phi} \\ \tilde{D}\,e^{i\phi} & \tilde{E} & -\tilde{D}\,e^{-i\phi} \\ \tilde{C}\,e^{i\phi} & -\tilde{B} & \tilde{A}\,e^{-i\phi} \end{bmatrix}_{c'c}, \tag{4.67}$$

where

$$\tilde{A} = \tfrac{1}{2}A(1 + \cos\theta_{\mathrm{lab}}) + \tfrac{1}{2}\sqrt{(2)}D\sin\theta_{\mathrm{lab}} + \tfrac{1}{2}C(1 - \cos\theta_{\mathrm{lab}}), \tag{4.68}$$

$$\tilde{B} = B\cos\theta_{\mathrm{lab}} + \tfrac{1}{2}\sqrt{(2)}E\sin\theta_{\mathrm{lab}}, \tag{4.69}$$

$$\tilde{C} = \tfrac{1}{2}C(1 + \cos\theta_{\mathrm{lab}}) - \tfrac{1}{2}\sqrt{(2)}D\sin\theta_{\mathrm{lab}} + \tfrac{1}{2}A(1 - \cos\theta_{\mathrm{lab}}), \tag{4.70}$$

$$\tilde{D} = -\tfrac{1}{2}\sqrt{(2)}A\sin\theta_{\mathrm{lab}} + D\cos\theta_{\mathrm{lab}} + \tfrac{1}{2}\sqrt{(2)}C\sin\theta_{\mathrm{lab}} = -\tilde{B}, \tag{4.71}$$

$$\tilde{E} = -\sqrt{(2)}B\sin\theta_{\mathrm{lab}} + E\cos\theta_{\mathrm{lab}} \tag{4.72}$$

$$I_0 = \tfrac{1}{3}(2|\tilde{A}|^2 + 2|\tilde{B}|^2 + 2|\tilde{C}|^2 + 2|\tilde{D}|^2 + |\tilde{E}|^2) \tag{4.73}$$

is identical with the I_0 of eqn (4.61) and X is given by eqn (3.42). The matrix $\tilde{M}_{c'c}$ transforms an initial density matrix ρ_c defined relative to some lab. axes (c) in which the z-axis is along the incident momentum into a resultant density matrix $\rho_{c'}$ ($= \tilde{M}_{c'c}\rho_c\tilde{M}_{c'c}^{\dagger}$) corresponding to a lab. frame c' with z'-axis along the outgoing momentum and y'-axis perpendicular to the scattering plane as in Section 3.4. This standard scattering matrix like $\hat{M}_{cc}$ has only four independent amplitudes.

4.4. Mueller method

As for spin-$\tfrac{1}{2}$ particles (eqn (3.66)) we may also describe the polarization state of a deuteron beam by a 'Stokes vector'

$$S_c = I \begin{bmatrix} 1 \\ t_{11} \\ t_{10} \\ t_{1\,-1} \\ t_{22} \\ t_{21} \\ t_{20} \\ t_{2\,-1} \\ t_{2\,-2} \end{bmatrix}_c, \tag{4.74}$$

in which the nine elements are essentially the intensity, vector, and tensor polarizations. Some of these elements are complex quantities, but again only nine real numbers are required to specify the polarization state, since some of the expectation values (e.g. t_{11} and $t_{1\,-1}$) are simply related.

The corresponding Mueller matrix $\tilde{Z}_{c'c}$ which operates on S_c to give a resultant Stokes vector

$$S_{c'} = \tilde{Z}_{c'c}S_c, \tag{4.75}$$

is 9×9 and may be obtained in terms of the 3×3 standard scattering matrix $\tilde{M}_{c'c}$ following the method of Section 1.8.1. The matrix $\tilde{Z}_{c'c}$ represents the effect of the elastic scattering on the polarization of the incident deuteron beam.

By analogy with eqn (3.79) for spin-$\frac{1}{2}$ particles, we may write for $\phi = 0$

$$\tilde{Z}_{c'c}(\phi = 0)$$

$$= X \begin{bmatrix} Z^{00}_{00} & Z^{00}_{11} & Z^{00}_{10} & Z^{00}_{1-1} & Z^{00}_{22} & Z^{00}_{21} & Z^{00}_{20} & Z^{00}_{2-1} & Z^{00}_{2-2} \\ Z^{11}_{00} & Z^{11}_{11} & Z^{11}_{10} & Z^{11}_{1-1} & Z^{11}_{22} & Z^{11}_{21} & Z^{11}_{20} & Z^{11}_{2-1} & Z^{11}_{2-2} \\ Z^{10}_{00} & Z^{10}_{11} & Z^{10}_{10} & Z^{10}_{1-1} & Z^{10}_{22} & Z^{10}_{21} & Z^{10}_{20} & Z^{10}_{2-1} & Z^{10}_{2-2} \\ Z^{1-1}_{00} & Z^{1-1}_{11} & Z^{1-1}_{10} & Z^{1-1}_{1-1} & Z^{1-1}_{22} & Z^{1-1}_{21} & Z^{1-1}_{20} & Z^{1-1}_{2-1} & Z^{1-1}_{2-2} \\ Z^{22}_{00} & Z^{22}_{11} & Z^{22}_{10} & Z^{22}_{1-1} & Z^{22}_{22} & Z^{22}_{21} & Z^{22}_{20} & Z^{22}_{2-1} & Z^{22}_{2-2} \\ Z^{21}_{00} & Z^{21}_{11} & Z^{21}_{10} & Z^{21}_{1-1} & Z^{21}_{22} & Z^{21}_{21} & Z^{21}_{20} & Z^{21}_{2-1} & Z^{21}_{2-2} \\ Z^{20}_{00} & Z^{20}_{11} & Z^{20}_{10} & Z^{20}_{1-1} & Z^{20}_{22} & Z^{20}_{21} & Z^{20}_{20} & Z^{20}_{2-1} & Z^{20}_{2-2} \\ Z^{2-1}_{00} & Z^{2-1}_{11} & Z^{2-1}_{10} & Z^{2-1}_{1-1} & Z^{2-1}_{22} & Z^{2-1}_{21} & Z^{2-1}_{20} & Z^{2-1}_{2-1} & Z^{2-1}_{2-2} \\ Z^{2-2}_{00} & Z^{2-2}_{11} & Z^{2-2}_{10} & Z^{2-2}_{1-1} & Z^{2-2}_{22} & Z^{2-2}_{21} & Z^{2-2}_{20} & Z^{2-2}_{2-1} & Z^{2-2}_{2-2} \end{bmatrix}_{c'c} ,$$

$$(4.76)$$

where the general element is of the form $Z^{K'Q'}_{KQ}$ and is given by (cf. eqns (3.85), (3.87), and (3.88) for spin-$\frac{1}{2}$ particles)

$$Z^{K'Q'}_{KQ} = \mathrm{tr}(\tilde{M}_{c'c}\hat{\tau}^{\dagger}_{KQ}\tilde{M}^{\dagger}_{c'c}\hat{\tau}_{K'Q'})/\mathrm{tr}(\tilde{M}_{c'c}\tilde{M}^{\dagger}_{c'c}). \qquad (4.77)$$

Again we have divided the elements (Mueller coefficients) into four groups: (1) $Z^{00}_{00} \equiv 1$, (2) Z^{00}_{1Q}, Z^{00}_{2Q}, (3)$Z^{1Q'}_{00}$, $Z^{2Q'}_{00}$, and (4) $Z^{K'Q'}_{KQ}$ with K, and $K' > 0$. The members of set (2) are related to the *vector* and *tensor analysing* powers T_{1Q} and T_{2Q}, respectively, as defined by the Madison convention (Barschall and Haeberli 1971) by the relationship

$$Z^{00}_{KQ} = (-1)^{Q}T_{K-Q} \qquad K = 1, 2. \qquad (4.78)$$

The elements of group (3) are referred to as the *vector* and *tensor polarizations* $t_{1Q'}$ and $t_{2Q'}$, respectively, *produced by an unpolarized incident beam*. The remaining coefficients are equivalent to the *polarization transfer coefficients* for $\phi = 0$ (Ohlsen 1972). The latter coefficients may be divided into four sub-groups according to the values of K and K': (1) vector–vector ($K = K' = 1$), (2) vector–tensor ($K = 1$, $K' = 2$), (3) tensor–vector ($K = 2$, $K' = 1$), and (4) tensor–tensor ($K = K' = 2$) polarization transfer coefficients, respectively.

4.4.1. *Parity conservation*

If the scattering process conserves parity, many of the Mueller coefficients of eqn (4.76) are simply related:

$$Z_{K-Q}^{K'-Q'} = (-1)^{K+K'+Q+Q'} Z_{KQ}^{K'Q'}. \tag{4.79}$$

Moreover, the coefficients are either real or pure imaginary since

$$(Z_{KQ}^{K'Q'})^* = (-1)^{K+K'} Z_{KQ}^{K'Q'}, \tag{4.80}$$

and in particular

$$Z_{K0}^{K'0} = 0 \qquad \text{for } (K+K') \text{ odd.} \tag{4.81}$$

The relations (4.79) and (4.80) are valid for any axes c, c' in which both the y-axis and the y'-axis are along $(\mathbf{k} \times \mathbf{k'})$, and may be proved by using eqn (4.51), which is valid for the elements of the standard scattering matrix $\tilde{M}_{c'c}$ for $\phi = 0$, in eqn (4.77), and noting that the elements of the operators $\hat{\tau}_{KQ}$ satisfy the relationship (see Section 5.1)

$$(\hat{\tau}_{KQ})_{\alpha\beta} = (-1)^{K+\alpha-\beta}(\hat{\tau}_{KQ}^\dagger)_{-\alpha-\beta}. \tag{4.82}$$

We have, omitting the reference axes,

$$
\begin{aligned}
(Z_{KQ}^{K'Q'})^* &= \{\text{tr}(\tilde{M}\hat{\tau}_{KQ}^\dagger \tilde{M}^\dagger \hat{\tau}_{K'Q'})\}^*/\text{tr}(\tilde{M}\tilde{M}^\dagger) \\
&= \sum_{\alpha\beta\gamma\delta} (\tilde{M})_{\alpha\beta}^*(\hat{\tau}_{KQ}^\dagger)_{\beta\gamma}^*(\tilde{M}^\dagger)_{\gamma\delta}^*(\hat{\tau}_{K'Q'})_{\delta\alpha}^*/\text{tr}(\tilde{M}\tilde{M}^\dagger) \\
&= \sum_{\alpha\beta\gamma\delta} (\tilde{M})_{\delta\gamma}(\hat{\tau}_{KQ})_{\gamma\beta}(\tilde{M}^\dagger)_{\beta\alpha}(\hat{\tau}_{K'Q'})_{\alpha\delta}/\text{tr}(\tilde{M}\tilde{M}^\dagger) \\
&= (-1)^{K+K'} \sum_{\alpha\beta\gamma\delta} (\tilde{M})_{-\delta-\gamma}(\hat{\tau}_{KQ}^\dagger)_{-\gamma-\beta}(\tilde{M}^\dagger)_{-\beta-\alpha}(\hat{\tau}_{K'Q'}^\dagger)_{-\alpha-\delta}/\text{tr}(\tilde{M}\tilde{M}^\dagger) \\
&= (-1)^{K+K'} Z_{KQ}^{K'Q'},
\end{aligned}
$$

and

$$
\begin{aligned}
Z_{K-Q}^{K'-Q'} &= \text{tr}(\tilde{M}\hat{\tau}_{K-Q}^\dagger \tilde{M}^\dagger \hat{\tau}_{K'-Q'})/\text{tr}(\tilde{M}\tilde{M}^\dagger) \\
&= (-1)^{Q+Q'} \sum_{\alpha\beta\gamma\delta} (\tilde{M})_{\alpha\beta}(\hat{\tau}_{KQ})_{\beta\gamma}(\tilde{M}^\dagger)_{\gamma\delta}(\hat{\tau}_{K'Q'}^\dagger)_{\delta\alpha}/\text{tr}(\tilde{M}\tilde{M}^\dagger) \\
&= (-1)^{K+K'+Q+Q'} \sum_{\alpha\beta\gamma\delta} (\tilde{M})_{\alpha\beta}^*(\hat{\tau}_{KQ})_{\beta\gamma}^*(\tilde{M}^\dagger)_{\gamma\delta}^*(\hat{\tau}_{K'Q'}^\dagger)_{\delta\alpha}^*/\text{tr}(\tilde{M}\tilde{M}^\dagger) \\
&= (-1)^{K+K'+Q+Q'} \sum_{\alpha\beta\gamma\delta} (\tilde{M})_{\delta\gamma}(\hat{\tau}_{KQ}^\dagger)_{\gamma\beta}(\tilde{M}^\dagger)_{\beta\alpha}(\hat{\tau}_{K'Q'})_{\alpha\delta}/\text{tr}(\tilde{M}\tilde{M}^\dagger) \\
&= (-1)^{K+K'+Q+Q'} Z_{KQ}^{K'Q'}.
\end{aligned}
$$

4.4.2. *Reciprocity relation*

If the axes c' associated with the scattered beam are taken to be in the c.m. system instead of the lab. frame, it is possible to relate each Mueller coefficient to the corresponding coefficient for the *inverse reaction*, which for elastic scattering is the same reaction. In this case, we have

$$Z_{K'Q'}^{KQ} = (-1)^{Q+Q'} Z_{KQ}^{K'Q'}.$$ (4.83)

In particular, if parity is also conserved,

$$Z_{00}^{KQ} = (-1)^{Q} Z_{KQ}^{00} = (-1)^{K} Z_{K-Q}^{00} \equiv (-1)^{K+Q} T_{KQ},$$ (4.84)

which for a given reaction relates the polarization produced for an unpolarized incident beam to the corresponding analysing power. Eqns (4.83)
and (4.84) also hold if the axes c, c' are identical with the z-axis along $(\mathbf{k}+\mathbf{k'})$
and the y-axis parallel to $(\mathbf{k} \times \mathbf{k'})$. The equations follow from the nature of
the scattering matrix, which for reciprocity has elements of the form (for
the axes chosen)

$$M_{\alpha\beta} = M_{\beta\alpha}(-1)^{\alpha-\beta},$$ (4.85)

and from observing that (see Section 5.1)

$$(\hat{\tau}_{KQ})_{\alpha\beta} = (-1)^{Q+\alpha-\beta}(\hat{\tau}_{KQ}^{\dagger})_{\beta\alpha}.$$ (4.86)

Thus

$$\begin{aligned}
Z_{KQ}^{K'Q'} &= \mathrm{tr}(M\hat{\tau}_{KQ}^{\dagger}M^{\dagger}\hat{\tau}_{K'Q'})/\mathrm{tr}(MM^{\dagger}) \\
&= \sum_{\alpha\beta\gamma\delta} (M)_{\alpha\beta}(\hat{\tau}_{KQ}^{\dagger})_{\beta\gamma}(M^{\dagger})_{\gamma\delta}(\hat{\tau}_{K'Q'})_{\delta\alpha}/\mathrm{tr}(MM^{\dagger}) \\
&= (-1)^{Q+Q'} \sum_{\alpha\beta\gamma\delta} (M)_{\beta\alpha}(\hat{\tau}_{K'Q'}^{\dagger})_{\alpha\delta}(M^{\dagger})_{\delta\gamma}(\hat{\tau}_{KQ})_{\gamma\beta}/\mathrm{tr}(MM^{\dagger}) \\
&= (-1)^{Q+Q'} Z_{K'Q'}^{KQ}.
\end{aligned}$$

4.4.3. Odd–even character of Mueller coefficients

Rotational invariance requires that the Mueller coefficients of eqn (4.76)
are odd or even functions of the scattering angle θ, according as to whether
the quantity $(Q+Q')$ is odd or even. This follows from invariance under a
rotation of 180° about the z-axis.

4.5. Double scattering of unpolarized deuterons

To indicate the usage of the Mueller formalism we consider the double
scattering of an unpolarized deuteron beam of unit intensity. Such a beam is
represented by the Stokes vector

$$S_c^{(0)} = \begin{bmatrix} 1 \\ 0 \\ \cdot \\ \cdot \\ \cdot \\ 0 \end{bmatrix}_c$$ (4.87)

where we assume the z-axis is along the direction of particle motion.

First scattering. For $\phi_1 = 0$, the scattered beam has the Stokes vector

$$S_{c'}^{(1)} = \tilde{Z}_{c'c}^{(1)} S_c^{(0)} = X^{(1)} \begin{bmatrix} 1 \\ \overset{(1)}{Z}_{00}^{11} \\ \overset{(1)}{Z}_{00}^{10} \\ \overset{(1)}{Z}_{00}^{1-1} \\ \overset{(1)}{Z}_{00}^{22} \\ \overset{(1)}{Z}_{00}^{21} \\ \overset{(1)}{Z}_{00}^{20} \\ \overset{(1)}{Z}_{00}^{2-1} \\ \overset{(1)}{Z}_{00}^{2-2} \end{bmatrix}_{c'} \equiv X^{(1)} \begin{bmatrix} 1 \\ t_{11}^{(1)} \\ 0 \\ t_{11}^{(1)} \\ t_{22}^{(1)} \\ t_{21}^{(1)} \\ t_{20}^{(1)} \\ -t_{21}^{(1)} \\ t_{22}^{(1)} \end{bmatrix}_{c'}, \qquad (4.88)$$

where the superscript (1) denotes first scattering and we have assumed the parity conservation relation (4.79) to obtain the last column. The scattered beam is described by five parameters, the intensity $X^{(1)}$, the vector polarization $t_{11}^{(1)}$ (pure imaginary), and the three real tensor polarizations $t_{20}^{(1)}$, $t_{21}^{(1)}$, $t_{22}^{(1)}$.

Second scattering. For arbitrary ϕ_2, the twice-scattered beam is given by

$$S_{c''}^{(2)} = \tilde{Z}_{c''c'}^{(2)} S_{c'}^{(1)}, \qquad (4.89)$$

where the elements of $\tilde{Z}_{c''c'}^{(2)}$ are related to the corresponding elements of $\tilde{Z}_{c''c'}^{(2)}(\phi_2 = 0)$, i.e. $Z_{KQ}^{(2)K'Q'}$, by a simple phase factor and have the general form

$$Z_{KQ}^{(2)K'Q'} e^{-iQ\phi_2}. \qquad (4.90)$$

Thus

$$(I^{(2)}t_{K'Q'}^{(2)})_{c''} = X^{(2)}X^{(1)} \sum_{KQ} Z_{KQ}^{(2)K'Q'} t_{KQ}^{(1)} e^{-iQ\phi_2}, \qquad (4.91)$$

and in particular, using eqns (4.78), (4.79), and (4.80), and replacing $(-T_{11}^{(2)}t_{11}^{(1)})$ by $(iT_{11}^{(2)})(it_{11}^{(1)})$, the intensity after double scattering is given in terms of real quantities:

$$I_{c''}^{(2)} = X^{(2)}X^{(1)}[(1 + T_{20}^{(2)}t_{20}^{(1)}) + 2\{(iT_{11}^{(2)})(it_{11}^{(1)}) + T_{21}^{(2)}t_{21}^{(1)}\}\cos\phi_2 +$$
$$+ 2T_{22}^{(2)}t_{22}^{(1)}\cos 2\phi_2]. \qquad (4.92)$$

By measuring $I_{c''}^{(2)}$ as a function of ϕ_2, information may be obtained about either the analysing powers $T_{KQ}^{(2)}$ or the polarizations $t_{KQ}^{(1)}$ which are functions of $(\theta_2)_{\text{lab}}$ and $(\theta_1)_{\text{lab}}$, respectively. In order to measure separately the various analysing powers, it is more convenient to employ beams of known polarization.

4.5.1. Scattering of a polarized beam

For an incident beam of polarization t_{KQ}^B, the intensity of the beam scattered along the direction $(\theta_{\text{lab}}, \phi)$ with respect to the incident direction is given (assuming the scattering conserves parity) by

$$I = X \sum_{KQ} t_{KQ}^B Z_{KQ}^{00} \, e^{-iQ\phi}$$

$$= X[1 + T_{20}t_{20}^B + 2\{(iT_{11})\,\text{Re}(it_{11}^B) + T_{21}\,\text{Re}(t_{21}^B)\}\cos\phi +$$

$$+ 2T_{22}\,\text{Re}(t_{22}^B)\cos 2\phi + 2\{(iT_{11})\,\text{Im}(it_{11}^B) + T_{21}\,\text{Im}(t_{21}^B)\}\sin\phi +$$

$$+ 2T_{22}\,\text{Im}(t_{22}^B)\sin 2\phi]. \tag{4.93}$$

To obtain this result, we have used eqns (4.79) and (4.81) and the relation

$$t_{K\,-Q}^B = (-1)^Q t_{KQ}^{B*}. \tag{4.94}$$

5

GENERAL FORMALISM

IN this chapter we generalize the non-relativistic formalism of Chapters 3 and 4 to include the interaction of particles of arbitrary spin with polarized targets.

5.1. Polarization of particles of arbitrary spin

Particles of spin s are associated with an intrinsic angular momentum operator $\hat{\mathbf{S}}$ which satisfies the commutation relations (2.23) with

$$\hat{\mathbf{S}}^2 = \hat{S}_x^2 + \hat{S}_y^2 + \hat{S}_z^2 = s(s+1)\hbar^2 \hat{\mathbf{1}}_{2s+1}. \tag{5.1}$$

The particles can exist in $(2s+1)$ basic spin states corresponding to the eigenvalues $m_s = -s\hbar, (-s+1)\hbar, \cdots, (s-1)\hbar, s\hbar$ of one of the components (say $\hat{S}_z$) of the operator $\hat{\mathbf{S}}$. A completely polarized beam may be represented by a $(2s+1)$-component spin wave function and is thus described by $(4s+1)$ independent quantities. An arbitrary beam is described by a $(2s+1) \times (2s+1)$ Hermitian density matrix $\rho(s)$ which has $(2s+1)^2$ independent quantities; the $(2s+1)$ real diagonal elements plus the real and imaginary parts of the $s(2s+1)$ independent off-diagonal elements.

For a complete representation of the density matrix $\rho(s)$ we may use a set of spherical tensor operators $\hat{\tau}_{KQ}(s)$ analogous to those of eqn (4.11) for deuterons. These $\hat{\tau}_{KQ}(s)$ may be represented by $(2s+1) \times (2s+1)$ matrices with elements (for a given coordinate system)

$$\{\hat{\tau}_{KQ}(s)\}_{\alpha\beta} = (2K+1)^{\frac{1}{2}} C(sKs, \beta Q\alpha), \tag{5.2}$$

where $C(j_1 j_2 j_3, m_1 m_2 m_3)$ are Clebsch–Gordan coefficients. Indeed, the spherical tensors of eqn (4.11) are of this form with $s = 1$ and eqns (4.12), (4.82), and (4.86) are readily proved using the symmetry relations for the Clebsch–Gordan coefficients (Rose 1957).

The expectation values of the operators $\hat{\tau}_{KQ}(s)$ are defined (analogous to eqn (4.17)) in terms of the density matrix (omitting the reference axes)

$$t_{KQ}(s) \equiv \langle \hat{\tau}_{KQ}(s) \rangle = \text{tr}\{\rho(s)\hat{\tau}_{KQ}(s)\}/\text{tr}\{\rho(s)\} \tag{5.3}$$

and, conversely, the density matrix which defines the state of beam polarization is given in terms of the expectation values (generalization of eqn (4.20))

$$\rho_c(s) = \frac{I}{(2s+1)} \left\{ \sum_{KQ} t_{KQ}(s)\hat{\tau}_{KQ}^\dagger(s) \right\}_c, \tag{5.4}$$

where $K = 0, 1, 2, \ldots, 2s$ and $Q = -K, -K+1, \ldots, K-1, K$. To specify the polarization completely all the $(2s+1)^2$ expectation values $t_{KQ}(s)$ must be known. For $K > 1$, the $t_{KQ}(s)$ will be referred to as *tensor polarizations of rank K*.

The *degree of polarization* is defined by (generalization of eqn (4.25))

$$D(s) = \left(\frac{1}{2s} \sum_{K=1}^{K=2s} \sum_Q |t_{KQ}(s)|^2 \right)^{\frac{1}{2}}. \tag{5.5}$$

For a completely polarized beam $D(s) = 1$; partially polarized beams have $D(s) < 1$ and unpolarized beams have $D(s) = 0$.

The polarization state of the beam may also be described by a Stokes vector (generalization of eqn (4.74))

$$S_c(s) = I \begin{bmatrix} 1 \\ t_{11} \\ \vdots \\ t_{2s \; -2s} \end{bmatrix}_c, \tag{5.6}$$

which has $(2s+1)^2$ elements. This representation also requires $(2s+1)^2$ real numbers to specify the polarization.

5.2. Elastic scattering of spin-s particles by a spinless target

The density matrix for an elastically scattered beam of spin-s particles is given in terms of the density matrix describing the incident beam $\rho_c^{(i)}(s)$ and a standard scattering matrix $\tilde{M}_{c'c}(s)$ which is a generalization of eqn (4.67) and describes the interaction for the standard coordinate axes c, c' defined in Section 4.3. We have

$$\rho_{c'}^{(f)}(s) = \tilde{M}_{c'c}(s)\rho_c^{(i)}(s)\tilde{M}_{c'c}^{\dagger}(s), \tag{5.7}$$

where all the quantities are $(2s+1) \times (2s+1)$ matrices.

By analogy with the expression (5.4) for the density matrix, the general form of the scattering matrix, assuming rotational invariance, may be written

$$\tilde{M}_{c'c}(s) = \frac{1}{(2s+1)} \sum_{KQ} \tilde{g}_{KQ}(s)\hat{t}_{KQ}^{\dagger}(s), \tag{5.8}$$

where the complex coefficients $\tilde{g}_{KQ}(s)$ are defined by

$$\tilde{g}_{KQ}(s) = \text{tr}\{\tilde{M}_{c'c}(s)\hat{t}_{KQ}(s)\}. \tag{5.9}$$

If parity conservation and reciprocity are required, the number of independent amplitudes $\tilde{g}_{KQ}(s)$ is considerably less than the total number of amplitudes $(2s+1)^2$.

5.2.1. *Parity conservation*

If parity is conserved in the scattering process, some of the coefficients $\tilde{g}_{KQ}(s)$ are not independent. For $\phi = 0$ we have

$$\tilde{g}_{K\,-Q}(s) = (-1)^{K+Q}\tilde{g}_{KQ}(s), \tag{5.10}$$

which may be proved by using eqn (4.82), and noting that the elements of $\tilde{M}_{c'c}(s)$ satisfy the relation

$$\{\tilde{M}_{c'c}(s)\}_{\alpha\beta} = (-1)^{\beta-\alpha}\{\tilde{M}_{c'c}(s)\}_{-\alpha\,-\beta}. \tag{5.11}$$

Relation (5.11) follows from eqn (11) of Berovic (1970) and from the equation

$$d^{(s)}_{\alpha\beta}(\theta) = (-1)^{\beta-\alpha}d^{(s)}_{-\alpha\,-\beta}(\theta) \tag{5.12}$$

for the elements of the reduced rotation matrix of eqn (4.16). Actually the condition given by eqn (5.10) is valid for *any* axes c, c' such that both the y and y' axes are along $(\mathbf{k}\times\mathbf{k}')$.

5.2.2. *Reciprocity relation*

If the axes c' associated with the scattered beam are taken to be in the c.m. system instead of the lab. frame, and the scattering process satisfies reciprocity, the coefficients $g_{KQ}(s)$ defining the corresponding scattering matrix $M_{c'c}(s)$ obey the constraint (dropping the tilde to denote the change from lab. to c.m. c' axes)

$$g_{K-Q}(s) = g_{KQ}(s). \tag{5.13}$$

This relation follows from a generalization of eqns (4.86) and (4.85), the latter equation in turn resulting from eqn (17) of Berovic (1970) with the appropriate rotation of axes.

5.2.3. *Number of independent elements of scattering matrix*

The scattering matrix $M_{c'c}(s)$ with z-axis along $\mathbf{k}$, z'-axis along $\mathbf{k}'$ (c.m.) and both the y- and y'-axes parallel to $(\mathbf{k}\times\mathbf{k}')$ subject to both parity conservation and reciprocity requirements is given essentially by eqns (5.8), (5.10), and (5.13). Thus for $s = \frac{1}{2}$ there are only two independent coefficients g_{00} and g_{11}, since $g_{10} = 0$ and $g_{1\,-1} = g_{11}$. For $s = 1$, parity conservation gives five independent amplitudes ($g_{00}, g_{11}, g_{20}, g_{21}$, and g_{22}) since $g_{10} = 0$, $g_{1\,-1} = g_{11}, g_{2\,-1} = -g_{21}$, and $g_{2\,-2} = g_{22}$; the requirement of reciprocity gives $g_{21} = 0$, reducing the number of independent coefficients to four. For arbitrary spin, parity conservation, and reciprocity lead to the following number of independent amplitudes:

$$\begin{aligned} n &= (s+1)^2 && \text{for bosons (integral spin)} \\ &= (s+\tfrac{1}{2})(s+\tfrac{3}{2}) && \text{for fermions (half-integral spin).} \end{aligned} \tag{5.14}$$

Neglecting an over-all phase factor, the number of independent combinations of the amplitudes which in principle are observable is $(2n-1)$. For example, for spin-$\frac{3}{2}$ particles, $n = 6$ and the number of independent observables is eleven. The results (5.14) hold for arbitrary axes c, c' although the simple relationships (5.10) and (5.13) are only valid for the special axes indicated.

5.2.4. Mueller method

The standard Mueller matrix $\tilde{Z}_{c'c}(s)$ which operates on an initial Stokes vector $S_c^{(i)}(s)$ to give a resultant Stokes vector

$$S_{c'}^{(f)}(s) = \tilde{Z}_{c'c}(s)S_c^{(i)}(s) \qquad (5.15)$$

is a $(2s+1)^2 \times (2s+1)^2$ matrix, which may be obtained in terms of the $(2s+1) \times (2s+1)$ standard scattering matrix $\tilde{M}_{c'c}$, following the method of Section 1.8.1.

By analogy with eqn (4.77) for deuterons, spin-s Mueller coefficients may be defined by the relation

$$Z_{KQ}^{K'Q'}(s) = \mathrm{tr}\{\tilde{M}_{c'c}(s)\hat{t}_{KQ}^{\dagger}(s)\tilde{M}_{c'c}^{\dagger}(s)\hat{t}_{K'Q'}(s)\}/\mathrm{tr}\{\tilde{M}_{c'c}(s)\tilde{M}_{c'c}^{\dagger}(s)\}, \qquad (5.16)$$

in which the azimuthal scattering angle is taken to be zero so that both the y- and y'-axes are along $(\mathbf{k} \times \mathbf{k}')$. If parity is conserved and reciprocity is obeyed, the results of Sections 4.4.1 and 4.4.2, respectively, are valid for the general Mueller coefficients $Z_{KQ}^{K'Q'}(s)$. The elements of the matrix $\tilde{Z}_{c'c}(s)$ for arbitrary ϕ are given by eqn (5.16) and are of the form (generalization of eqn (4.90))

$$Z_{KQ}^{K'Q'}(s)\,\mathrm{e}^{-\mathrm{i}Q\phi}. \qquad (5.17)$$

5.3. Elastic scattering of spin-s particles by targets with spin

We now consider target particles which possess non-zero spin t, and discuss both the elastic scattering process and, in the next section, the more general reaction in which the two final particles may be different from the initial pair of particles.

5.3.1. Polarization of initial system

Consider a beam of particles with spin s incident upon a target composed of particles (for the present assumed to be distinguishable from the incident particles) with spin t. A completely polarized (i.e. pure) state of the incident beam *and* target system in which the particles are non-interacting may be represented by a $(2s+1)(2t+1)$-component wave function which is the direct product of a $(2s+1)$-component wave function describing the completely polarized beam and a $(2t+1)$-component wave function for a completely polarized target, i.e.

$$\chi_c(s,t) = \chi_c(s) \otimes \chi_c(t), \qquad (5.18)$$

where c denotes the reference axes, and the direct product $\otimes$ has the meaning indicated in eqn (1.59). Expanding $\chi_c(s)$ and $\chi_c(t)$ in terms of diagonalized basis functions $\phi_c^{(m_s)}$, $\phi_c^{(m_t)}$, respectively, we can write

$$\chi_c(s, t) = \sum_{m_s m_t} a_{m_s} a_{m_t} (\phi_c^{(m_s)} \otimes \phi_c^{(m_t)})$$

$$= \sum_{m_s m_t} C_{m_s m_t} \phi_c^{(m_s, m_t)}, \qquad (5.19)$$

where the functions $\phi_c^{(m_s, m_t)}$ of the product spin-space form a diagonalized basis and are represented by $(2s+1)(2t+1)$-column matrices with every element zero except the (m_s, m_t) element, which is unity. Thus we can write

$$\chi_c(s, t) = \begin{bmatrix} C_{s\,t} \\ C_{s\,t-1} \\ \vdots \\ C_{s\,-t} \\ C_{s-1\,t} \\ \vdots \\ C_{-s\,-t} \end{bmatrix}_c, \qquad (5.20)$$

where $|C_{m_s m_t}|^2$ gives the probability of finding the incident beam in spin state m_s and the target in spin state m_t, simultaneously. Such a pure state of the system is described by $(4s+4t+1)$ independent quantities.

An arbitrary (incident-beam plus target) system may be represented by a $(2s+1)(2t+1) \times (2s+1)(2t+1)$ Hermitian density matrix $\rho_c(s, t)$ which may be written as the direct product of the density matrices $\rho_c(s)$ and $\rho_c(t)$ describing the polarization states of the incident beam and target, respectively. In general, the matrix $\rho_c(s, t)$ has $\{(2s+1)^2 + (2t+1)^2 - 1\}$ independent quantities and for its complete representation we may use the direct product of spherical tensor operators $\hat{\tau}_{KQ}(s)$ and $\hat{\tau}_{kq}(t)$ associated with the respective spin sub-spaces s and t:

$$\hat{\tau}_{KQkq}(s, t) = \{\hat{\tau}_{KQ}(s) \otimes \hat{\tau}_{kq}(t)\}, \qquad (5.21)$$

and denote the corresponding matrix elements by a four-index system:

$$\{\hat{\tau}_{KQkq}(s, t)\}_{\alpha\gamma\beta\delta} = \{\hat{\tau}_{KQ}(s)\}_{\alpha\beta} \{\hat{\tau}_{kq}(t)\}_{\gamma\delta}$$

$$= (2K+1)^{\frac{1}{2}} (2k+1)^{\frac{1}{2}} C(sKs, \beta Q\alpha) C(tkt, \delta q\gamma). \qquad (5.22)$$

The ordering of these elements within the matrix $\hat{\tau}_{KQkq}(s, t)$ is such that the later indices vary more rapidly, analogously to the elements of the direct product of eqn (1.59). For example,

$$\hat{\tau}_{1111}(1,\tfrac{1}{2}) = \hat{\tau}_{11}(1)\otimes\hat{\tau}_{11}(\tfrac{1}{2})$$

$$= -\sqrt{\left(\tfrac{3}{2}\right)}\begin{bmatrix} 0 & 1 & 0 \\ 0 & 0 & 1 \\ 0 & 0 & 0 \end{bmatrix} \otimes -\sqrt{2}\begin{bmatrix} 0 & 1 \\ 0 & 0 \end{bmatrix}$$

column indices

$$= \sqrt{3}\;\begin{array}{c c}
\begin{array}{cccccc} 1\tfrac{1}{2} & 1-\tfrac{1}{2} & 0\tfrac{1}{2} & 0-\tfrac{1}{2} & -1\tfrac{1}{2} & -1-\tfrac{1}{2} \end{array} & \\
\left[\begin{array}{cc|cc|cc}
0 & 0 & 0 & 1 & 0 & 0 \\ \hline
0 & 0 & 0 & 0 & 0 & 0 \\ \hline
0 & 0 & 0 & 0 & 0 & 1 \\ \hline
0 & 0 & 0 & 0 & 0 & 0 \\ \hline
0 & 0 & 0 & 0 & 0 & 0 \\ \hline
0 & 0 & 0 & 0 & 0 & 0
\end{array}\right] &
\begin{array}{l}
1\tfrac{1}{2} \\
1-\tfrac{1}{2} \\
0\tfrac{1}{2} \\
0-\tfrac{1}{2} \\
-1\tfrac{1}{2} \\
-1-\tfrac{1}{2}
\end{array}
\end{array}$$

(row indices)

The expectation values of the operators $\hat{\tau}_{KQkq}(s, t)$ are defined by

$$t_{KQkq}(s, t) \equiv \langle \hat{\tau}_{KQkq}(s, t)\rangle = \mathrm{tr}\{\rho_c(s, t)\hat{\tau}_{KQkq}(s, t)\}/\mathrm{tr}\{\rho_c(s, t)\}$$

$$= t_{KQ}(s)t_{kq}(t), \qquad (5.23)$$

i.e. equal to the product of the expectation values corresponding to the incident beam and target individually. $\rho_c(s, t)$ may be written

$$\rho_c(s, t) = \frac{\mathrm{tr}\{\rho_c(s, t)\}}{(2s+1)(2t+1)}\sum_{KQkq} t_{KQkq}(s, t)\hat{\tau}^{\dagger}_{KQkq}(s, t), \qquad (5.24)$$

where it is conventional to normalize $\mathrm{tr}\{\rho_c(t)\}$ to unity so that $\mathrm{tr}\{\rho_c(s, t)\} \equiv I$, the intensity of the incident beam.

The degree of polarization of the system may be defined by the quantity

$$D(s, t) = \frac{1}{\sqrt{\{(2s+1)(2t+1)-1\}}}\left\{\sum_{KQkq} |t_{KQkq}(s, t)|^2 - 1\right\}^{\tfrac{1}{2}}, \qquad (5.25)$$

so that a completely polarized system has $D(s, t) = 1$, a partially polarized system has $D(s, t) < 1$ and an unpolarized system has $D(s, t) = 0$.

The polarization state of the system may also be described by a straightforward generalization of the Stokes vector of eqn (5.6):

$$
S_c(s, t) = I \begin{bmatrix} 1 \\ \vdots \\ t_{KQkq}(s, t) \\ \vdots \end{bmatrix}_c \equiv I \begin{bmatrix} 1 \\ \vdots \\ t_{KQ}(s)t_{kq}(t) \\ \vdots \end{bmatrix}_c ,
\tag{5.26}
$$

i.e. a column vector with $(2s+1)^2(2t+1)^2$ elements and $\{(2s+1)^2 + (2t+1)^2 - 1\}$ independent real quantities.

5.3.2. *Standard scattering matrix*

The density matrix for an elastically scattered beam of spin-s particles from a spin-t target is given by the product of $(2s+1)(2t+1) \times (2s+1)(2t+1)$ matrices:

$$
\rho_c^{(f)}(s, t) = \tilde{M}_{c'c}(s, t)\rho_c^{(i)}(s, t)\tilde{M}_{c'c}^{\dagger}(s, t),
\tag{5.27}
$$

where $\rho_c^{(i)}(s, t)$ is the density matrix describing the initial system and $\tilde{M}_{c'c}(s, t)$ is the standard scattering matrix for the system, which by a straightforward generalization of eqns (3.53) and (4.67) satisfies the relation

$$
\tilde{M}_{c'c}(s, t) = (X/I_0)^{\frac{1}{2}}\{D^{(s)}(\phi, \theta_{\text{lab}}, 0)^{\dagger} \otimes D^{(t)}(\phi, \theta_{\text{lab}}, 0)^{\dagger}\}\hat{M}_{cc}(s, t),
\tag{5.28}
$$

where $\hat{M}_{cc}$ is the scattering matrix for axes c and

$$
I_0 = \frac{1}{(2s+1)(2t+1)} \text{tr}\{\hat{M}_{cc}(s, t)\hat{M}_{cc}^{\dagger}(s, t)\}.
\tag{5.29}
$$

The quantity (X/I_0) is defined by eqn (3.42) in which $d\Omega_{\text{lab}}$ is the lab. solid angle for the final beam of particles s. It should be noted that the intensity of the final beam of particles t is in general different from X by the ratio of the lab. solid angles.

By analogy with eqn (5.24) and generalizing eqn (5.8), the standard scattering matrix may be expanded in terms of direct products of spherical tensor operators:

$$
\tilde{M}_{c'c}(s, t) = \frac{1}{(2s+1)(2t+1)} \sum_{KQkq} \tilde{g}_{KQkq}(s, t)\hat{t}_{KQkq}^{\dagger}(s, t),
\tag{5.30}
$$

where

$$
\tilde{g}_{KQkq}(s, t) = \text{tr}\{\tilde{M}_{c'c}(s, t)\hat{t}_{KQkq}(s, t)\}.
\tag{5.31}
$$

For $\phi = 0$, the coefficients $\tilde{g}_{KQkq}(s, t)$ satisfy the relation

$$
\tilde{g}_{K-Qk-q}(s, t) = (-1)^{K+Q+k+q}\tilde{g}_{KQkq}(s, t),
\tag{5.32}
$$

if *parity is conserved*. Thus under this condition

$$\tilde{g}_{K0k0}(s, t) = 0 \quad \text{for } (K+k) \text{ odd}. \tag{5.33}$$

As for eqn (5.10), the parity relation (5.32) is valid for any axes c, c' where both the y- and y'-axes are along $(\mathbf{k} \times \mathbf{k}')$.

For axes c' in the c.m. frame we have a reciprocity relation analogous to eqn (5.13), i.e.

$$g_{K-Qk-q}(s, t) = g_{KQkq}(s, t) \tag{5.34}$$

if the scattering process satisfies *reciprocity*.

Using the relations (5.32) and (5.34), the number of independent amplitudes n of the general rotational-invariant scattering matrix $M_{c'c}(s, t)$, which satisfies parity conservation and reciprocity requirements, may be obtained (cf. Section 5.2.3). The results are

(1) reciprocity alone:

$$n = (2s+1)(2t+1)(2st+s+t+1); \tag{5.35}$$

(2) parity conservation alone:

$$n = \tfrac{1}{2}(2s+1)^2(2t+1)^2 \qquad \text{if } either \, s \text{ or } t \text{ half-integral}$$
$$= \tfrac{1}{2}(2s+1)^2(2t+1)^2 + \tfrac{1}{2} \qquad \text{if } both \, s \text{ and } t \text{ integral}; \tag{5.36}$$

(3) reciprocity and parity conservation:

$$n = \tfrac{1}{4}(2s+1)(2t+1)(4st+2s+2t+3) \qquad \text{if } either \, s \text{ or } t \text{ half-integral}$$
$$= (2st+s+t+1)^2 \qquad \text{if } both \, s \text{ and } t \text{ integral}. \tag{5.37}$$

Thus, for example, if $s = \tfrac{1}{2}$, $t = 1$, the scattering matrix has 21, 18, and 12 independent amplitudes under the conditions (1), (2), and (3), respectively.

5.3.3. Mueller matrix

The standard Mueller matrix $\tilde{Z}_{c'c}(s, t)$ which transforms an initial Stokes vector $S_c^{(i)}(s, t)$ into the Stokes vector for the scattered particles

$$S_c^{(f)}(s, t) = \tilde{Z}_{c'c}(s, t)S_c^{(i)}(s, t) \tag{5.38}$$

is a $(2s+1)^2(2t+1)^2 \times (2s+1)^2(2t+1)^2$ matrix corresponding to the standard scattering matrix $\tilde{M}_{c'c}(s, t)$ in the usual manner. For zero azimuthal scattering angle we can define Mueller coefficients, which are a generalization of eqn (5.16):

$$Z_{KQkq}^{K'Q'k'q'}(s, t) = \frac{\text{tr}\{\tilde{M}_{c'c}(s, t)\hat{\tau}_{KQkq}^{\dagger}(s, t)\tilde{M}_{c'c}^{\dagger}(s, t)\hat{\tau}_{K'Q'k'q'}(s, t)\}}{\text{tr}\{\tilde{M}_{c'c}(s, t)\tilde{M}_{c'c}^{\dagger}(s, t)\}}. \tag{5.39}$$

If parity is conserved we have (analogous to eqn (4.79))

$$Z_{K-Qk-q}^{K'-Q'k'-q'}(s, t) = (-1)^{K+K'+k+k'+Q+Q'+q+q'} Z_{KQkq}^{K'Q'k'q'}(s, t). \tag{5.40}$$

Rotational invariance requires that the Mueller coefficients of eqn (5.39) are odd or even functions of the scattering angle θ according to whether the quantity $(Q+Q'+q+q')$ is odd or even. If reciprocity is obeyed and the axes c' are in the c.m. frame, then (cf. eqn (4.83))

$$Z_{K'Q'k'q'}^{KQkq}(s, t) = (-1)^{Q+Q'+q+q'} Z_{KQkq}^{K'Q'k'q'}(s, t). \qquad (5.41)$$

The elements of $\tilde{Z}_{c'c}$ for arbitrary ϕ are related to the corresponding elements of $\tilde{Z}_{c'c}(\phi = 0)$ by a simple phase factor and have the general form

$$Z_{KQkq}^{K'Q'k'q'}(s, t)\, e^{-i(Q+q)\phi}. \qquad (5.42)$$

Thus for the standard axes we have the general result:

$$\{I^{(f)}t_{K'Q'k'q'}(s, t)\}_{c'} = X \sum_{KQkq} Z_{KQkq}^{K'Q'k'q'}(s, t)\{I^{(i)}t_{KQkq}(s, t)\}_c\, e^{-i(Q+q)\phi}$$

$$= X \sum_{KQkq} Z_{KQkq}^{K'Q'k'q'}(s, t)\{I^{(i)}t_{KQ}(s)t_{kq}(t)\}_c\, e^{-i(Q+q)\phi}. \qquad (5.43)$$

The relation

$$t_{KQkq}(s, t) = t_{KQ}(s)t_{kq}(t), \qquad (5.44)$$

which we have used in eqn (5.43), is only valid for particles s and t which are *not correlated* in any way, e.g. as in the initial system. For the final system, unless either $K' = Q' = 0$ or $k' = q' = 0$,

$$t_{K'Q'k'q'}(s, t) \neq t_{K'Q'}(s)t_{k'q'}(t), \qquad (5.45)$$

since in general the density matrix $\rho_{c'}^{(f)}(s, t)$ of eqn (5.27) cannot be expressed as a direct product of $\rho_{c'}^{(f)}(s)$ and $\rho_{c'}^{(f)}(t)$ corresponding to an uncorrelated final system.

5.4. Non-elastic reactions involving two final particles

We now consider the rather general case in which two distinguishable particles with spins s and t respectively interact to produce two resultant distinguishable particles with spins s' and t' and which may be different from the initial pair of particles, i.e. the reaction

$$s+t \rightarrow s'+t'. \qquad (5.46)$$

5.4.1. Standard reaction matrix

The density matrix for the resultant system is given by (generalizing eqn (5.27))

$$\rho_{c'}^{(f)}(s', t') = \tilde{M}_{c'c}(s', t'; s, t)\rho_c^{(i)}(s, t)\tilde{M}_{c'c}^\dagger(s', t'; s, t), \qquad (5.47)$$

where $\rho_c^{(i)}(s, t)$ is the density matrix (eqn (5.24)) describing the initial system and $\tilde{M}_{c'c}(s', t'; s, t)$ is the standard *reaction* matrix for the process (5.46), which is analogous to the standard scattering matrix $\tilde{M}_{c'c}(s, t; s, t) \equiv \tilde{M}_{c'c}(s, t)$ of eqn (5.28), and satisfies the relation

$$\tilde{M}_{c'c}(s', t'; s, t) = (X/I_0)^{\frac{1}{2}} \{ D^{(s')}(\phi, \theta_{\text{lab}}, 0)^\dagger \otimes D^{(t')}(\phi, \theta_{\text{lab}}, 0)^\dagger \} \hat{M}_{cc}(s', t'; s, t),$$

(5.48)

where $\hat{M}_{cc}(s', t'; s, t)$ is the reaction matrix for axes c and

$$I_0 = \frac{1}{(2s+1)(2t+1)} \, \text{tr}\{ \hat{M}_{cc}(s', t'; s, t) \hat{M}^\dagger_{cc}(s', t'; s, t)\}. \qquad (5.49)$$

The standard reaction matrix is a $(2s'+1)(2t'+1) \times (2s+1)(2t+1)$ rectangular matrix which relates the initial and final density matrices. The initial system is referred to lab. axes c in which the z-axis is along the direction of motion of the incident beam of spin-s particles, and the final system is defined relative to lab. axes c' in which the z'-axis is along the direction of motion of the outgoing spin-s' particles and the y'-axis is normal to the reaction plane and parallel to $(\mathbf{k} \times \mathbf{k}'(s'))$, where $\mathbf{k}'(s')$ is the momentum of the spin-s' particles.

The standard reaction matrix may be expanded in terms of direct products of 'operators' $\hat{\tau}'_{KQkq}(s, t; s', t')$, i.e.

$$\tilde{M}_{c'c}(s', t'; s, t) = \frac{1}{(2s+1)(2t+1)} \sum_{KQkq} \tilde{g}_{KQkq}(s', t'; s, t) \hat{\tau}'^\dagger_{KQkq}(s, t; s', t'), \qquad (5.50)$$

where

$$\tilde{g}_{KQkq}(s', t'; s, t) = \text{tr}\{ \tilde{M}_{c'c}(s', t'; s, t) \hat{\tau}'_{KQkq}(s, t; s', t')\} \qquad (5.51)$$

and

$$\hat{\tau}'_{KQkq}(s, t; s', t') = \hat{\tau}'_{KQ}(s, s') \otimes \hat{\tau}'_{kq}(t, t'), \qquad (5.52)$$

and has elements

$$\{ \hat{\tau}'_{KQkq}(s, t; s', t')\}_{\alpha\gamma\beta\delta} = (2K+1)^{\frac{1}{2}} (2k+1)^{\frac{1}{2}} C(s'Ks, \beta Q\alpha) C(t'kt, \delta q\gamma). \qquad (5.53)$$

The quantities $\hat{\tau}'_{KQ}(s, s')$ and $\hat{\tau}'_{kq}(t, t')$ are not spherical tensor operators but are generalizations of the latter operators, in the sense that their matrix elements are Clebsch–Gordan coefficients of the form

$$\{ \hat{\tau}'_{KQ}(s, s')\}_{\alpha\beta} = (2K+1)^{\frac{1}{2}} C(s'Ks, \beta Q\alpha), \qquad (5.54)$$

where s' is generally not equal to s.

If *parity is conserved* the coefficients $\tilde{g}_{KQkq}(s', t'; s, t)$ satisfy the relation (cf. eqn (5.32))

$$\tilde{g}_{K-Qk-q}(s', t'; s, t) = \eta(-1)^{K+Q+k+q} \tilde{g}_{KQkq}(s', t'; s, t), \qquad (5.55)$$

for $\phi = 0$. Here $\eta = -1$ if there is a change of intrinsic parities between the initial and final channels, and $\eta = +1$ otherwise. Eqn (5.55) may be verified using the relation (Berovic (1970))

$$\{\hat{M}_{cc}(s', t'; s, t)\}_{\alpha\beta\gamma\delta} = \eta(-1)^{s'+t'-s-t+\beta+\delta-\alpha-\gamma}\{\hat{M}_{cc}(s', t'; s, t)\}_{-\alpha -\beta -\gamma -\delta}$$

$$(5.56)$$

with eqn (5.12) for the elements of the rotation matrix $D^{(K)}(0, \theta_{\text{lab}}, 0)$ and the symmetry relations of the Clebsch–Gordan coefficients.

Relation (5.55), which holds for any axes c, c' such that both the y- and y'-axes are parallel to $(\mathbf{k} \times \mathbf{k}'(s'))$, is equivalent to eqn (1) of Bohr (1959) defined for different axes. For the latter axes in which the z-axis is perpendicular to the reaction plane, eqn (5.55) becomes essentially the *Bohr theorem*:

$$g_{KQkq}(s', t'; s, t) = \eta(-1)^{Q+q}g_{KQkq}(s', t'; s, t),\qquad (5.57)$$

i.e.

$$g_{KQkq}(s', t'; s, t) = 0$$

unless $(Q+q) \equiv \{(\alpha+\gamma)-(\beta+\delta)\}$ is even (odd) for $\eta = +1(-1)$. Here α, β, γ, and δ are the spin projections for particles s', s, t', and t, respectively, along the z-axis.

5.4.2. Mueller matrix

The standard Mueller matrix $\tilde{Z}_{c'c}(s', t'; s, t)$, which transforms an initial Stokes vector $S_c^{(i)}(s, t)$ into the Stokes vector for the resultant particles,

$$S_{c'}^{(f)}(s', t') = \tilde{Z}_{c'c}(s', t'; s, t)S_c^{(i)}(s, t),\qquad (5.58)$$

is a $(2s'+1)^2(2t'+1)^2 \times (2s+1)^2(2t+1)^2$ rectangular matrix corresponding to the standard reaction matrix $\tilde{M}_{c'c}(s', t'; s, t)$. For zero azimuthal reaction angle (for particle s') we have a relation corresponding to eqn (5.39) for elastic scattering:

$$Z_{KQkq}^{K'Q'k'q'}(s', t'; s, t) = \frac{\text{tr}\{\tilde{M}_{c'c}(s', t'; s, t)\hat{\tau}_{KQkq}^{\dagger}(s, t)\tilde{M}_{c'c}^{\dagger}(s', t'; s, t)\hat{\tau}_{K'Q'k'q'}(s', t')\}}{\text{tr}\{\tilde{M}_{c'c}(s', t'; s, t)\tilde{M}_{c'c}^{\dagger}(s', t'; s, t)\}}.$$

$$(5.59)$$

If parity is conserved in the reaction we have a relation analogous to eqn (5.40)

$$Z_{K-Qk-q}^{K'-Q'k'-q'}(s', t'; s, t) = (-1)^{K+K'+k+k'+Q+Q'+q+q'}Z_{KQkq}^{K'Q'k'q'}(s', t'; s, t).\quad (5.60)$$

Rotational invariance requires that the Mueller coefficients of eqn (5.59) are odd or even functions of the reaction angle θ (particle s') according as the quantity $(Q+Q'+q+q')$ is odd or even. The elements of $\tilde{Z}_{c'c}$ for arbitrary azimuthal reaction angle ϕ (particle s') are related to the corresponding elements of $\tilde{Z}_{c'c}(\phi = 0)$ by a simple phase factor and have the general form

$$Z_{KQkq}^{K'Q'k'q'}(s', t'; s, t)\,e^{-i(Q+q)\phi}, \tag{5.61}$$

and consequently for the standard axes we have the general result analogous to eqn (5.43) for elastic scattering

$$\{I^{(f)}t_{K'Q'k'q'}(s', t')\}_{c'} = X \sum_{KQkq} Z_{KQkq}^{K'Q'k'q'}(s', t'; s, t)\{I^{(i)}t_{KQ}(s)t_{kq}(t)\}_c\,e^{-i(Q+q)\phi} \tag{5.62}$$

The elements $Z_{KQkq}^{K'Q'k'q'}(s', t'; s, t)$ of the Mueller matrix $\tilde{Z}_{c'c}(\phi = 0)$ may be conveniently divided into twelve groups:

(1) $Z_{0000}^{0000} \equiv 1$;

(2) $Z_{KQ00}^{0000} = (-1)^Q T_{K\,-Q}(s)$ where $T_{KQ}(s)$ are the *analysing powers* for initial particle s;

(3) $Z_{00kq}^{0000} = (-1)^q T_{k\,-q}(t)$, where $T_{kq}(t)$ are the *analysing powers* for initial particle t;

(4) $Z_{0000}^{K'Q'00} = t_{K'Q'}^{(0)}(s')$, the *polarizations* of final particle s' produced by an unpolarized initial system;

(5) $Z_{0000}^{00k'q'} = t_{k'q'}^{(0)}(t')$, the *polarizations* of final particle t' produced by an unpolarized initial system;

(6) $Z_{KQkq}^{0000} = \mathscr{S}_{KQkq}(s, t)$, the *spin-correlation coefficients* for the initial channel;

(7) $Z_{0000}^{K'Q'k'q'} = \mathscr{S}^{K'Q'k'q'}(s', t')$, the *spin-correlation coefficients* for the final channel;

(8) $Z_{KQ00}^{K'Q'00} = \mathscr{O}_{KQ}^{K'Q'}(s', s)$, the *polarization transfer coefficients* for initial particle s to final particle s';

(9) $Z_{00kq}^{00k'q'} = \mathscr{O}_{kq}^{k'q'}(t', t)$, the *polarization transfer coefficients* for initial particle t to final particle t';

(10) $Z_{00kq}^{K'Q'00} = \mathscr{O}_{kq}^{K'Q'}(s', t)$, the *polarization transfer coefficients* for initial particle t to final particle s';

(11) $Z_{KQ00}^{00k'q'} = \mathscr{O}_{KQ}^{k'q'}(t', s)$, the *polarization transfer coefficients* for initial particle s to final particle t';

(12) $Z_{KQkq}^{K'Q'k'q'}$;

where in all cases except (12), K, K', k, and k' are non-zero unless explicitly specified.

5.4.3. *Relation to inverse reaction*

Although the reciprocity relations analogous to eqns (5.34) and (5.41) do not hold for non-elastic reactions, since the reaction and its inverse process are different, the amplitudes $g_{K\,-Qk\,-q}(s', t'; s, t)$ and the elements $Z_{K'Q'k'q'}^{KQkq}(s', t'; s, t)$ are related to the corresponding quantities for the *inverse reaction* if reciprocity is obeyed. We have (for the axes c' in the c.m. frame)

$$g_{K\,-Qk\,-q}(s', t'; s, t) \propto g_{KQkq}(s, t; s', t'), \tag{5.63}$$

where the constant of proportionality depends solely upon kinematic factors, and

$$Z^{KQkq}_{K'Q'k'q'}(s', t'; s, t) = (-1)^{Q+Q'+q+q'} Z^{K'Q'k'q'}_{KQkq}(s, t; s', t').$$ (5.64)

Note carefully the ordering of the arguments s, t, s', t', since for non-elastic reactions

$$Z^{K'Q'k'q'}_{KQkq}(s', t'; s, t) \neq Z^{K'Q'k'q'}_{KQkq}(s, t; s', t'),$$ (5.65)

for example, even if the particles s' and t' have the same spin values as the particles s and t, respectively. Our prime notation simply implies that the reaction is non-elastic (and allows possible different spin values).

5.5. Arrow notation and the classes of experiments

It has become conventional (Barschall and Haeberli 1971) to place an arrow over the symbol which denotes a particle which is either in a polarized state or whose state of polarization is measured. Thus for the reaction

$$\vec{t}(\vec{s}, s')t' \quad \text{(spin-correlation experiment)},$$ (5.66)

the arrows signify that the final intensity is measured for a polarized beam s incident upon a polarized target t. Similarly,

$$t(\vec{s}, \vec{s}')t' \quad \text{(polarization transfer experiment)}$$ (5.67)

indicates that the polarization of the final beam s' is measured for a polarized beam s incident upon an unpolarized target t.

We now discuss some of the various classes of experiments within the framework of eqn (5.62).

5.5.1. Polarization experiments

Experiments of the kind $t(s, \vec{s}')t'$ in which the initial system is unpolarized and the polarization of one of the final particles is measured are called *polarization experiments*. For the standard axes we have from eqn (5.62)

$$\{I^{(f)} t_{K'Q'}(s')\}_{c'} = XZ^{K'Q'00}_{0000} \{I^{(i)}\}_c,$$

i.e.

$$t_{K'Q'}(s') = Z^{K'Q'00}_{0000} \equiv t^{(0)}_{K'Q'}(s'),$$ (5.68)

since $I^{(f)} = XI^{(i)}$. If the polarizations $t_{K'Q'}(s')$ are measured by some appropriate analysing reaction, the coefficients $Z^{K'Q'00}_{0000}$ may be determined.

5.5.2. Analysing power experiments

Experiments in which either the incident beam s or the target t is polarized and only the final intensity is measured are called *analysing power experiments*. For a typical reaction $t(\vec{s}, s')t'$ we have

$$\{I^{(f)}\}_{c'} = X \sum_{KQ} Z^{0000}_{KQ00} \{I^{(i)} t_{KQ}(s)\}_c \, e^{-iQ\phi}$$

$$\equiv X \sum_{KQ} (-1)^Q T_{K\,-Q} \{I^{(i)} t_{KQ}(s)\}_c \, e^{-iQ\phi}, \qquad (5.69)$$

and using known polarizations $t_{KQ}(s)$ the analysing powers T_{KQ} may be measured.

5.5.3. *Spin-correlation experiments*

There are two kinds of *spin-correlation experiments*, (a) $\vec{t}(\vec{s}, s')t'$ and (b) $t(s, \vec{s}')\vec{t}'$, which are useful for measuring the initial and final channel spin-correlation coefficients $\mathscr{S}_{KQkq}(s, t)$ and $\mathscr{S}^{K'Q'k'q'}(s', t')$, respectively. In case (a) only the final intensity is measured, and we have

$$\{I^{(f)}\}_{c'} = X \sum_{KQkq} Z^{0000}_{KQkq} \{I^{(i)} t_{KQkq}(s, t)\}_c \, e^{-i(Q+q)\phi}$$

$$= X I^{(i)} \Bigg[1 + \sum_{K>0,Q} (-1)^Q T_{K\,-Q}(s) \{t_{KQ}(s)\}_c \, e^{-iQ\phi} +$$

$$+ \sum_{k>0,q} (-1)^q T_{k\,-q}(t) \{t_{kq}(t)\}_c \, e^{-iq\phi} +$$

$$+ \sum_{K>0, k>0, Q, q} \mathscr{S}_{KQkq}(s, t) \{t_{KQ}(s) t_{kq}(t)\}_c \, e^{-i(Q+q)\phi} \Bigg], \qquad (5.70)$$

and provided the analysing powers $T_{KQ}(s)$ and $T_{kq}(t)$ have been measured in simpler experiments, the spin-correlation coefficients $\mathscr{S}_{KQkq}(s, t)$ may be obtained by employing appropriate polarizations $t_{KQ}(s)$ and $t_{kq}(t)$.

In case (b) the simultaneous spin orientations of both final particles, i.e. $t_{K'Q'k'q'}(s', t')$, are measured which essentially determines the spin-correlation coefficients $\mathscr{S}^{K'Q'k,q'}(s', t')$, since we have

$$\{I^{(f)} t_{K'Q'k'q'}(s', t')\}_{c'} = X Z^{K'Q'k'q'}_{0000} \{I^{(i)}\}_c,$$

i.e.

$$t_{K'Q'k'q'}(s', t') = Z^{K'Q'k'q'}_{0000} \equiv \mathscr{S}^{K'Q'k'q'}(s', t'), \qquad (5.71)$$

since $I^{(f)} = X I^{(i)}$.

5.5.4. *Polarization transfer experiments*

Reactions of the type (5.67) in which one of the colliding particles is polarized and the polarization of one of the final particles is measured are called *polarization transfer experiments*. For the reaction $t(\vec{s}, \vec{s}')t'$ we have

$$\{I^{(f)}t_{K'Q'}(s')\}_{c'} = X \sum_{KQ} Z_{KQ00}^{K'Q'00} \{I^{(i)}t_{KQ}(s)\}_c\, e^{-iQ\phi}$$

$$= XI^{(i)}\left[t_{K'Q'}^{(0)}(s') + \sum_{K>0,Q} \mathcal{O}_{KQ}^{K'Q'}(s',s)\{t_{KQ}(s)\}_c\, e^{-iQ\phi} \right], \qquad (5.72)$$

and using known polarizations $t_{KQ}(s)$, the polarization transfer coefficients $\mathcal{O}_{KQ}^{K'Q'}(s',s)$ may be determined by measuring $t_{K'Q'}(s')$ and $t_{K'Q'}^{(0)}(s')$.

5.6. Expansion of Mueller coefficients in reaction matrix amplitudes

The Mueller coefficients of eqn (5.59) may be expressed in terms of the reaction matrix amplitudes $\tilde{g}_{KQkq}$ of eqn (5.51). Substituting the expansion (5.50) for $\tilde{M}_{c'c}$ and using the relation (5.53) for the quantities $\hat{\tau}'_{KQkq}$ on the right-hand side of eqn (5.59) we obtain

$$Z_{KQkq}^{K'Q'k'q'}(s',t';s,t) = \sum_{\tilde{K}\tilde{Q}\tilde{k}\tilde{q}\tilde{K}'\tilde{Q}'\tilde{k}'\tilde{q}'} \tilde{g}_{\tilde{K}\tilde{Q}\tilde{k}\tilde{q}} \tilde{g}_{\tilde{K}'\tilde{Q}'\tilde{k}'\tilde{q}'}^* B_{KQkq\tilde{K}\tilde{Q}\tilde{k}\tilde{q}}^{K'Q'k'q'\tilde{K}'\tilde{Q}'\tilde{k}'\tilde{q}'}(s',t';s,t), \qquad (5.73)$$

where

$$B_{KQkq\tilde{K}\tilde{Q}\tilde{k}\tilde{q}}^{K'Q'k'q'\tilde{K}'\tilde{Q}'\tilde{k}'\tilde{q}'}(s',t';s,t)$$
$$= [(2K+1)(2k+1)(2K'+1)(2k'+1)(2\tilde{K}+1)(2\tilde{k}+1)(2\tilde{K}'+1)(2\tilde{k}'+1)]^{\frac{1}{2}} \times$$
$$\times \sum_{\alpha\beta\gamma\delta\varepsilon\zeta\eta\theta} C(s'\tilde{K}s, \alpha\tilde{Q}\beta)C(t'\tilde{k}t, \gamma\tilde{q}\delta)C(sKs, \beta Q\varepsilon)C(tkt, \delta q\zeta) \times$$
$$\times C(s'\tilde{K}'s, \eta\tilde{Q}'\varepsilon)C(t'\tilde{k}'t, \theta\tilde{q}'\zeta)C(s'K's', \alpha Q'\eta)C(t'k't', \gamma q'\theta). \qquad (5.74)$$

Summing over the magnetic quantum numbers gives

$$B_{KQkq\tilde{K}\tilde{Q}\tilde{k}\tilde{q}}^{K'Q'k'q'\tilde{K}'\tilde{Q}'\tilde{k}'\tilde{q}'}(s',t';s,t)$$
$$= \{(2s+1)(2t+1)(2s'+1)(2t'+1)(2K'+1)(2k'+1)(2\tilde{K}+1)(2\tilde{k}+1)\}^{\frac{1}{2}} \times$$
$$\times \sum_{Jj} (2J+1)(2j+1)X(\tilde{K}\tilde{K}'J, s's'K', ssK)X(\tilde{k}\tilde{k}'j, t't'k', ttk) \times$$
$$\times C(JK'K, Q'-Q, -Q'-Q)C(\tilde{K}J\tilde{K}', \tilde{Q}Q-Q'\tilde{Q}')C(jk'k, q'-q, -q'-q) \times$$
$$\times C(\tilde{k}j\tilde{k}', \tilde{q}q-q'\tilde{q}'), \qquad (5.75)$$

where $X(j_1j_2j_3, j_4j_5j_6, j_7j_8j_9)$ are the nine j- or X-coefficients as defined by Brink and Satchler (1968). In particular,

$$B_{0000\tilde{K}\tilde{Q}\tilde{k}\tilde{q}}^{0000\tilde{K}'\tilde{Q}'\tilde{k}'\tilde{q}'}(s',t';s,t) = \delta_{\tilde{K}\tilde{K}'}\,\delta_{\tilde{Q}\tilde{Q}'}\,\delta_{\tilde{k}\tilde{k}'}\,\delta_{\tilde{q}\tilde{q}'}. \qquad (5.76)$$

The above procedure may be extended readily to reactions in which an arbitrary number of particles interact (see Csonka, Moravcsik, and Scadron (1966)).

5.7. Identical particles

We now consider the cases in which either the initial pair of particles or the final pair of particles (or both pairs) are *indistinguishable*. For such identical particles we must take into account the fact that only properly symmetrized pure states are allowed and that the mixed states permitted are incoherent superpositions of these symmetrized pure states.

The complete wave functions describing the possible pure states of two identical particles must be symmetric or antisymmetric under the particle-exchange operator, depending whether the particles are bosons or fermions. Thus, for a pair of fermions, if the two-particle spin wave function is symmetric, the corresponding spatial wave function is required to be antisymmetric, so that the complete wave function is antisymmetric. Unfortunately, the direct-product spin wave functions which we have used previously and which are very convenient for describing the polarizations of the individual particles do not possess a definite symmetry for particle exchange. However, if we couple the spins (t, say) together to form states of total spin S and projection M, the resultant coupled spin wave functions

$$\chi_{SM} = \sum_{m_1 m_2} C(ttS, m_1 m_2 M)(\chi_{tm_1} \otimes \chi_{tm_2}) \tag{5.77}$$

do have a definite symmetry. If $\hat{P}_{12}$ is the particle-exchange operator we have

$$\begin{aligned}
\hat{P}_{12}\chi_{SM} &= \sum_{m_1 m_2} C(ttS, m_1 m_2 M)\hat{P}_{12}(\chi_{tm_1} \otimes \chi_{tm_2}) \\
&= \sum_{m_1 m_2} C(ttS, m_1 m_2 M)(\chi_{tm_2} \otimes \chi_{tm_1}) \\
&= \sum_{m_1 m_2} (-1)^{2t-S} C(ttS, m_2 m_1 M)(\chi_{tm_2} \otimes \chi_{tm_1}) \\
&= (-1)^{2t-S}\chi_{SM}, \tag{5.78}
\end{aligned}$$

i.e. χ_{SM} is symmetric or antisymmetric according as the quantity $(2t-S)$ is even or odd. The corresponding spatial wave function (for both bosons and fermions) is therefore symmetric or antisymmetric according as S is even or odd. Thus by expressing the standard reaction matrix in terms of the coupled spin representation it is straightforward to generalize to the case of identical particles.

For distinguishable particles we have

$$\begin{aligned}
&\{\tilde{M}_{c'c}(s', t'; s, t)\}_{\alpha\gamma\beta\delta} \\
&\qquad = \sum_{S'M'SM} C(s't'S', \alpha\gamma M')C(stS, \beta\delta M)\{\tilde{M}_{c'c}(s', t'; s, t)\}_{S'M'SM}. \tag{5.79}
\end{aligned}$$

Now the reaction matrix is a function of the c.m. incoming and outgoing relative momenta $\mathbf{k}$ and $\mathbf{k}'$, i.e.

$$\tilde{M}_{c'c}(s', t'; s, t) \equiv \tilde{M}_{c'c}(\mathbf{k}', \mathbf{k}), \quad \text{say}, \tag{5.80}$$

If the initial pair of particles are identical we have to replace $\{\tilde{M}_{c'c}\}_{S'M'SM}$ in eqn (5.79) by the appropriate symmetrized matrix element involving both $\mathbf{k}$ and $-\mathbf{k}$ contributions, i.e. (dropping the axes)

$$\{\tilde{M}(\mathbf{k}', \mathbf{k})\}_{S'M'SM} \rightarrow \frac{1}{\sqrt{2}}\{\tilde{M}(\mathbf{k}', \mathbf{k})+(-1)^S\tilde{M}(\mathbf{k}', -\mathbf{k})\}_{S'M'SM}. \quad (5.81)$$

Similarly, for identical final particles

$$\{\tilde{M}(\mathbf{k}', \mathbf{k})\}_{S'M'SM} \rightarrow \frac{1}{\sqrt{2}}\{\tilde{M}(\mathbf{k}', \mathbf{k})+(-1)^{S'}M(-\mathbf{k}', \mathbf{k})\}_{S'M'SM}, \quad (5.82)$$

and for both pairs indistinguishable

$$\{\tilde{M}(\mathbf{k}', \mathbf{k})\}_{S'M'SM} \rightarrow \tfrac{1}{2}\{\tilde{M}(\mathbf{k}', \mathbf{k})+(-1)^S\tilde{M}(\mathbf{k}', -\mathbf{k})+$$
$$+(-1)^{S'}\tilde{M}(-\mathbf{k}', \mathbf{k})+(-1)^{S+S'}\tilde{M}(-\mathbf{k}', -\mathbf{k})\}_{S'M'SM}. \quad (5.83)$$

5.7.1. *Special condition*

For identical particles, the number of independent amplitudes (n) of the scattering matrix given by eqn (5.37) for parity conservation and reciprocity is modified. For identical particles the additional condition

$$g_{KQkq}(s, s) = g_{kqKQ}(s, s) \quad (5.84)$$

holds, and one finds for particles of spin s that

$$n = \tfrac{1}{6}(s+1)(2s+1)(6s^2+5s+6). \quad (5.85)$$

Thus for proton–proton elastic scattering ($s = \frac{1}{2}$) we have only *five* independent amplitudes compared with six for distinguishable particles as given by eqn (5.37). For deuteron–deuteron elastic scattering ($s = 1$), eqn (5.85) gives $n = 17$.

6

EMISSION AND ABSORPTION OF ELECTROMAGNETIC RADIATION

WHEN a system in a definite quantum state decays to another pure state by the emission of energy in the form of electromagnetic radiation, the resultant radiation is polarized. In this chapter we describe the polarization of such radiation and also the inverse process in which polarized radiation is used to obtain a non-random distribution of the quantized orientations of a system possessing spin (e.g. optical pumping for atomic systems). The methods developed in the previous chapter will be seen to apply to these cases, with the additional requirement in some instances that the final states are separated into components (cf. calcite crystal). We now consider one of the first methods of studying decaying pure quantum states, i.e. Zeeman's use of an external magnetic field to split the degenerate energy levels of atomic systems.

6.1. Zeeman effect

In a magnetic field, the normal energy levels of an atom are split into several sub-levels corresponding to the discrete orientations of atomic magnetic moments. If the energy level has spin j there are $(2j+1)$ sub-levels with 'magnetic' quantum numbers $m = -j, -j+1, \dots, j-1, j$; and energies $E(m) = E_0 + \Delta E(m)$, where E_0 is the energy of the degenerate levels in the absence of a field. Thus a magnetic field changes a normal spectral line with frequency

$$v = (E_0^{(i)} - E_0^{(f)})/h \tag{6.1}$$

into several spectral lines (or components) with frequencies

$$v(m_f, m_i) = [\{E_0^{(i)} + \Delta E(m_i)\} - \{E_0^{(f)} + \Delta E(m_f)\}]/h$$

$$= v + \{\Delta E(m_i) - \Delta E(m_f)\}/h. \tag{6.2}$$

Zeeman discovered that the number of components and their relative intensities and polarizations depend upon the direction of the field relative to the observer. For example, when viewed perpendicular to the field the outer (σ) components are linearly polarized perpendicular to the field while the inner (π) components are linearly polarized parallel to the field. In the longitudinal direction (i.e. parallel to the field) the σ-components are circularly polarized while the π-components are absent altogether. These observations may be understood as a consequence of angular momentum and parity conservation by considering the radiation field as a complete set of multipole fields.

6.2. Multipole fields

The electromagnetic field may be expanded in terms of a complete set of spherical standing waves (multipoles) corresponding to fields of definite angular momentum J and parity π_y (see Rose (1955) for a general discussion). For large r (measured from source), the photons of these multipole fields are described by wave functions $\psi(JM\pi_y)$ given by

$$\psi(JM\pi_y) \simeq \left\{ \sum_{lm\mu} b_{Jl} C(1lJ, \mu m M) Y_{lm}(\theta, \phi) \chi_\mu \right\} \times \text{function of } r, \qquad (6.3)$$

where the spin wave functions χ_μ $(\mu = 0, \pm 1)$ are defined by eqn (2.39), and the coefficients b_{Jl} required to satisfy the *transversality condition* for electromagnetic radiation are of the form (n.b. J and $l \geqslant 0$)

$$b_{J\,J-1} = 0, \qquad\qquad b_{JJ} = 1, \qquad b_{J\,J+1} = 0 \qquad\qquad \text{for } \pi_y = (-1)^{J+1}$$

$$= \left(\frac{J+1}{2J+1}\right)^{\frac{1}{2}}, \qquad\quad = 0, \qquad\quad = \left(\frac{J}{2J+1}\right)^{\frac{1}{2}} \qquad = (-1)^J.$$

$$(6.4)$$

For each angular momentum J there are two multipole fields according to the parity π_y: *electric* multipole $(\mathscr{E}J)$ with $\pi_y = (-1)^J$ and *magnetic* multipole $(\mathscr{M}J)$ with $\pi_y = (-1)^{J+1}$. The multipoles are called dipole, quadrupole, octupole, etc. for $J = 1, 2, 3, \dots$, respectively. There is no electric monopole field $(\mathscr{E}0)$ corresponding to $J = 0$ and $l = 1$ since the coefficient $b_{01} \equiv 0$. Likewise, there is no magnetic monopole field $(\mathscr{M}0)$ since $C(100, \mu m M) \equiv 0$ for $J = l = 0$.

In the radiative decay of a system from an initial pure state (j_i, m_i, π_i) to another pure state (j_f, m_f, π_f), where j, m, and π are the spin, spin projection, and parity, respectively, only those photons associated with multipole fields which conserve angular momentum and parity for the whole system are emitted, i.e.

$$j_{\min} \leqslant J \leqslant j_i + j_f \quad \text{and} \quad \pi_y = \pi_i/\pi_f, \qquad (6.5)$$

where $j_{\min}$ is the greater of 1 or $|j_i - j_f|$. Thus for $j_i = j_f = \frac{1}{2}$, $\pi_i = -1$ and $\pi_f = +1$, we have $J = 1$, $\pi_y = -1$ only, i.e. electric dipole $(\mathscr{E}1)$ radiation: for $j_i = j_f = 0$ radiative decay is forbidden.

6.3. Decay matrix

Consider the *spontaneous* radiative decay of a system j_i:

$$j_i \to j_f + \gamma, \qquad (6.6)$$

where again we use the same symbol to denote both the system and its spin. As for the reaction $(s + t \to s' + t')$, we may describe the density matrix for the resultant system in terms of a matrix operator $\hat{\Delta}_{c'c}(j_f, \gamma; j_i)$ and the density matrix $\rho_c^{(i)}(j_i)$ for the initial system, i.e.

$$\rho_c^{(f)}(j_f, \gamma) = \hat{\Delta}_{c'c}(j_f, \gamma; j_i)\rho_c^{(i)}(j_i)\hat{\Delta}_{c'c}^{\dagger}(j_f, \gamma; j_i). \tag{6.7}$$

The $3(2j_f + 1) \times (2j_i + 1)$ matrix $\hat{\Delta}_{c'c}(j_f, \gamma; j_i)$ is analogous to the reaction matrix $\tilde{M}_{c'c}$ $(s', t'; s, t)$ of eqn (5.50) and is termed the *decay matrix*. We shall use the same axes for the initial and final states and define a 'standard decay matrix' $\bar{\Delta}_{cc} \equiv \hat{\Delta}_{cc}$. Where possible this notation will also imply axes with the z-axis along the axis of symmetry (e.g. the magnetic field direction).

If the initial system is in a pure state described by the wave function $\chi_c^{(i)}$ the resultant system is also in a pure state $\chi_c^{(f)}$ given by (cf. eqn (3.2))

$$\chi_c^{(f)} = \bar{\Delta}_{cc}\chi_c^{(i)}. \tag{6.8}$$

The matrix elements of $\bar{\Delta}_{cc}$ are of the form

$$(\bar{\Delta}_{cc})_{m_f\mu\, m_i} = \sum_{JM} a(j_f Jj_i)C(j_f Jj_i, m_f M m_i) \sum_{lm} b_{Jl}C(1lJ, \mu m M)Y_{lm}^*(\theta, \phi), \tag{6.9}$$

where $a(j_f Jj_i)$ is the reduced matrix element (or 'nuclear factor') which depends upon the detailed structure of the initial and final states but is independent of the magnetic quantum numbers if the interaction causing the transition satisfies rotational invariance and $C(j_f Jj_i, m_f M m_i)$ is the 'geometrical factor' arising from the Wigner–Eckart theorem (see e.g. Brink and Satchler (1968)). In eqn (6.9) the Clebsch–Gordan coefficients are zero unless $M = (m_i - m_f)$ and $m = (m_i - m_f - \mu)$, respectively.

6.3.1. *Zeeman effect on mercury (even isotope) 2537 Å line*

As an example let us consider the effect of a magnetic field upon the mercury 2537 Å line. For even isotopes, which have zero nuclear spin, this line results from the decay of the 6^3P_1 state to the 6^1S_0 state ($n^{2s+1}L_J$ is the conventional spectroscopic notation for a level with n as the principal quantum number and S, P, D, ... corresponding to orbital angular momentum $L = 0, 1, 2, ...$). In a magnetic field, the line is split into three components (see Fig. 6.1) corresponding to the three possible transitions between the three initial sub-levels and the singlet state (we ignore any possible transitions between components of the 6^3P_1 state which would correspond to $\mathcal{M}1$ or $\mathcal{E}2$ radiation). The reader should consult an appropriate text, e.g. Shore and Menzel (1968), for a discussion of the various multipole transition rates: in general however $\mathcal{E}1$ radiation dominates atomic spectroscopy. For the three transitions shown the emitted light is $\mathcal{E}1$ radiation and the elements of the appropriate decay matrix are omitting the reduced matrix element

$$(\bar{\Delta}_{cc})_{m_f\mu\, m_i} \propto C(011, m_f M m_i)\{\sqrt{\tfrac{2}{3}}C(101, \mu m M)Y_{0m}^* +$$

$$+ \sqrt{\tfrac{1}{3}}C(121, \mu m M)Y_{2m}^*\}, \tag{6.10}$$

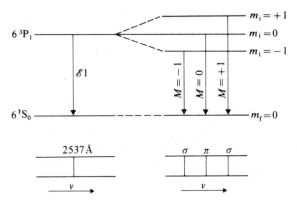

FIG. 6.1. Zeeman effect for $6^3P_1 \rightarrow 6^1S_0(\mathscr{E}1)$ transition in mercury (even isotope). In a magnetic field the normal 2537 Å line is separated into one $\pi(M = 0)$ and two $\sigma(M = \pm 1)$ components.

giving

$$\bar{\Delta}_{cc} \propto \sqrt{\left(\frac{3}{8\pi}\right)} \times$$

$$\times \begin{bmatrix} \frac{1}{2}(1+\cos^2\theta) & \frac{1}{\sqrt{2}}\sin\theta\cos\theta\,e^{i\phi} & \frac{1}{2}\sin^2\theta\,e^{2i\phi} \\ \frac{1}{\sqrt{2}}\sin\theta\cos\theta\,e^{-i\phi} & \sin^2\theta & -\frac{1}{\sqrt{2}}\sin\theta\cos\theta\,e^{i\phi} \\ \frac{1}{2}\sin^2\theta\,e^{-2i\phi} & -\frac{1}{\sqrt{2}}\sin\theta\cos\theta\,e^{-i\phi} & \frac{1}{2}(1+\cos^2\theta) \end{bmatrix}_{cc},$$

$$(6.11)$$

where the z-axis is along the field direction.

Thus for the decay of the state $m_i = +1$ we have for unit initial 'population'

$$\chi_c^{(f)} = \bar{\Delta}_{cc} \begin{bmatrix} 1 \\ 0 \\ 0 \end{bmatrix}_c \propto (1)_c \otimes \begin{bmatrix} \frac{1}{2}(1+\cos^2\theta) \\ \frac{1}{\sqrt{2}}\sin\theta\cos\theta\,e^{-i\phi} \\ \frac{1}{2}\sin^2\theta\,e^{-2i\phi} \end{bmatrix}_c = (1)_c \otimes \chi_c^{(y)}(1\rightarrow 0), \quad \text{say}, \quad (6.12)$$

where we have written the final spin wave function as a direct product of a scalar wave function describing the final atomic state and a wave function $\chi_c^{(y)}(m_i \rightarrow m_f)$ describing the emitted radiation. The latter wave functions are represented in terms of the diagonalized basis states of eqn (2.39). In order to transform to a representation using the non-diagonalized Cartesian basis states of eqn (2.25) corresponding essentially to the Jones vectors of Chapters

1 and 2, it is necessary to multiply the wave functions $\chi_c^{(\gamma)}$ by the unitary matrix

$$U = \frac{1}{\sqrt{2}} \begin{bmatrix} -1 & 0 & 1 \\ -i & 0 & -i \\ 0 & \sqrt{2} & 0 \end{bmatrix}. \tag{6.13}$$

In the new basis we obtain

$$\chi_c^{(\gamma)}(1 \to 0) = -\frac{1}{2\sqrt{2}} \begin{bmatrix} (1+\cos^2\theta) - \sin^2\theta\, e^{-2i\phi} \\ i(1+\cos^2\theta) + i\sin^2\theta\, e^{-2i\phi} \\ -2\sin\theta\cos\theta\, e^{-i\phi} \end{bmatrix}_c. \tag{6.14}$$

Thus for $\theta = 0$ we have

$$\chi_c^{(\gamma)}(1 \to 0) = -\frac{1}{\sqrt{2}} \begin{bmatrix} 1 \\ i \\ 0 \end{bmatrix}_c, \tag{6.15}$$

i.e. right-handed circularly polarized light. For $\theta = \tfrac{1}{2}\pi$ and $\phi = 0$ we have

$$\chi_c^{(\gamma)}(1 \to 0) = -\frac{1}{\sqrt{2}} \begin{bmatrix} 0 \\ i \\ 0 \end{bmatrix}_c, \tag{6.16}$$

corresponding to light linearly polarized along the y-axis (i.e. perpendicular to the field).

Similarly, in the new basis, the other two components resulting from the decay of the $m_i = 0$ and $m_i = -1$ states have the respective wave functions

$$\chi_c^{(\gamma)}(0 \to 0) = \begin{bmatrix} -\sin\theta\cos\theta\cos\phi \\ \sin\theta\cos\theta\sin\phi \\ \sin^2\theta \end{bmatrix}_c \tag{6.17}$$

and

$$\chi_c^{(\gamma)}(-1 \to 0) = \frac{1}{2\sqrt{2}} \begin{bmatrix} -\sin^2\theta\, e^{2i\phi} + (1+\cos^2\theta) \\ -i\sin^2\theta\, e^{2i\phi} - i(1+\cos^2\theta) \\ -2\sin\theta\cos\theta\, e^{i\phi} \end{bmatrix}_c. \tag{6.18}$$

Thus for $\theta = 0$

$$\chi_c^{(\gamma)}(0 \to 0) = \begin{bmatrix} 0 \\ 0 \\ 0 \end{bmatrix}_c \quad \text{and} \quad \chi_c^{(\gamma)}(-1 \to 0) = \frac{1}{\sqrt{2}} \begin{bmatrix} 1 \\ -i \\ 0 \end{bmatrix}_c, \tag{6.19}$$

and the components are null and left-handed circularly polarized, respectively. For $\theta = \pi/2$ and $\phi = 0$

$$\chi_c^{(\gamma)}(0 \to 0) = \begin{bmatrix} 0 \\ 0 \\ 1 \end{bmatrix}_c \quad \text{and} \quad \chi_c^{(\gamma)}(-1 \to 0) = \frac{1}{\sqrt{2}} \begin{bmatrix} 0 \\ -i \\ 0 \end{bmatrix}_c, \quad (6.20)$$

i.e. the emitted radiations are linearly polarized along the z-axis (i.e. parallel to the field) and along the y-axis (i.e. perpendicular to the field), respectively. These results are in agreement with the observations as described in Section 6.1.

The relative intensities of the three components are given by

$$I(m_i \to m_f) \propto \chi_c^{(\gamma)\dagger}(m_i \to m_f)\chi_c^{(\gamma)}(m_i \to m_f), \quad (6.21)$$

and we have, neglecting the very small differences in the relative populations $N(m_i)$ of the $m_i = 0, \pm 1$ states arising from the Boltzmann distribution corresponding to temperature T (k here is Boltzmann's constant),

$$\frac{N(m_i = \pm 1)}{N(m_i = 0)} = \exp[\{\Delta E(m_i = \pm 1) - \Delta E(m_i = 0)\}/kT], \quad (6.22)$$

that

$$I(0 \to 0) \equiv I_\pi = \tfrac{1}{2}I_n \sin^2 \theta, \quad (6.23)$$

and

$$I(+1 \to 0) = I(-1 \to 0) \equiv I_\sigma = \tfrac{1}{4}I_n(1 + \cos^2 \theta), \quad (6.24)$$

where

$$I_n = \sum_{m_i} I(m_i \to 0) \quad (6.25)$$

is the intensity of the normal line, i.e. in the absence of a field.

6.4. Mueller method

Instead of using the density matrix description, the initial and final states j_i and (j_f, γ) of the decay (6.6) may be represented by the Stokes vectors

$$S_c^{(i)}(j_i) = I^{(i)} \begin{bmatrix} 1 \\ \vdots \\ t_{KQ}(j_i) \\ \vdots \end{bmatrix}_c \quad \text{and} \quad S_c^{(f)}(j_f, \gamma) = I^{(f)} \begin{bmatrix} 1 \\ \vdots \\ t_{K'Q'k'q'}(j_f, \gamma) \\ \vdots \end{bmatrix}_c, \quad (6.26)$$

respectively, where

$$t_{KQ}(j_i) = \text{tr}\{\rho_c^{(i)}(j_i)\hat{\tau}_{KQ}(j_i)\}/\text{tr}\{\rho_c^{(i)}(j_i)\} \quad (6.27)$$

and

$$t_{K'Q'k'q'}(j_f, \gamma) = \mathrm{tr}[\rho_c^{(f)}(j_f, \gamma)\{\hat{\tau}_{K'Q'}(j_f) \otimes \hat{\tau}_{k'q'}(\gamma)\}]/\mathrm{tr}\{\rho_c^{(f)}(j_f, \gamma)\} \quad (6.28)$$

are the expectation values of the operators $\hat{\tau}_{KQ}(j_i)$ and $\hat{\tau}_{K'Q'k'q'}(j_f, \gamma)$ representing the initial and final systems. We have

$$S_c^{(f)}(j_f, \gamma) = \bar{Z}_{cc}(j_f, \gamma; j_i)S_c^{(i)}(j_i), \quad (6.29)$$

where (omitting the axes) $\bar{Z}$ is the 'standard' Mueller matrix corresponding to $\bar{\Delta}$ of eqn (6.9) and has Mueller coefficients given by

$$Z_{KQ}^{K'Q'k'q'}(j_f, \gamma; j_i) = \mathrm{tr}\{\bar{\Delta}\hat{\tau}_{KQ}^\dagger(j_i)\bar{\Delta}^\dagger\hat{\tau}_{K'Q'k'q'}(j_f, \gamma)\}/\mathrm{tr}(\bar{\Delta}\bar{\Delta}^\dagger), \quad (6.30)$$

in which the azimuthal angle ϕ of the emitted photon is taken to be zero so that the y-axis is along $\mathbf{f} \times \mathbf{k'}$, where $\mathbf{f}$ is a unit vector along the symmetry axis (if any) and $\mathbf{k'}$ is the momentum of the resultant photon.

Under *parity conservation* (assumed for eqn (6.9) for $\bar{\Delta}$) it is readily shown that the Mueller coefficients satisfy the usual parity relation (of eqn (5.60)):

$$Z_{K\ -Q}^{K'\ -Q'k'\ -q'}(j_f, \gamma; j_i) = (-1)^{K+K'+k'+Q+Q'+q'}Z_{KQ}^{K'Q'k'q'}(j_f, \gamma; j_i). \quad (6.31)$$

Rotational invariance requires that the Mueller coefficients are odd or even functions of the photon decay angle θ according as the quantity $(Q+Q'+q')$ is odd or even. This is seen by using eqn (6.9) (which assumes rotational invariance) for $\bar{\Delta}$ with $\phi = 0$ and θ replaced by $-\theta$, and eqns (5.2) and (5.22) for the spherical tensor operators in eqn (6.30).

The Mueller decay matrix defined by eqn (6.30), like the Mueller reaction matrix $\bar{Z}$ $(s', t'; s, t)$ of eqn (5.59), assumes that no separation is made of the various final states (j_f, γ) corresponding to definite values of either of the spin projections m_f or μ. If this were so (e.g. m_f in the Zeeman effect if $j_f \neq 0$) the appropriate Mueller matrices describing the decays to states of definite spin projection may still be obtained from eqn (6.30), provided the elements of $\bar{\Delta}$ corresponding to unobserved final spin projections are set equal to zero (cf. stopped calcite crystal of Section 1.5).

The elements of $\bar{Z}_{cc}$ for arbitrary azimuthal photon angle ϕ are related to the corresponding Mueller coefficients by a simple phase factor and have the general form

$$Z_{KQ}^{K'Q'k'q'}(j_f, \gamma; j_i)\, e^{-i(Q-Q'-q')\phi}. \quad (6.32)$$

Thus the general element of the final Stokes vector $S_c^{(f)}$ is given by

$$\{I^{(f)}t_{K'Q'k'q'}(j_f, \gamma)\}_c$$

$$= I_0 \sum_{KQ} Z_{KQ}^{K'Q'k'q'}(j_f, \gamma, j_i)\{I^{(i)}t_{KQ}(j_i)\}_c\, e^{-i(Q-Q'-q')\phi}, \quad (6.33)$$

where (cf. eqn (5.49))

$$I_0 = \frac{1}{(2j_i+1)} \mathrm{tr}\{\bar{\Delta}_{cc}(j_f, \gamma; j_i)\bar{\Delta}_{cc}^\dagger(j_f, \gamma; j_i)\}. \quad (6.34)$$

6.5. Gamma correlation experiments

As examples of the Mueller method for decay we consider both gamma–gamma and particle–gamma correlation experiments.

6.5.1. *Gamma–gamma correlation*

In a gamma–gamma correlation experiment one gamma ray γ_1 emitted by a radioactive nucleus is observed in some fixed direction $\mathbf{k}_1$, and the angular distribution of a succeeding gamma ray γ_2 with respect to the direction $\mathbf{k}_1$ is measured. Without the presence of an external field the spin orientations of a radioactive nucleus are normally isotropic: the observation of γ_1 along some fixed direction serves to select out a non-isotropic distribution of spin orientations, so that in general the succeeding γ_2 shows an *angular correlation* with respect to the direction of γ_1. We shall only discuss unperturbed correlations, i.e. cases in which the intermediate state is not changed before emission of the second radiation. Let us consider the successive decays

$$j_0 \rightarrow j_1 + \gamma_1, \qquad j_1 \rightarrow j_2 + \gamma_2. \tag{6.35}$$

First decay. In general the initial state j_0 is unpolarized and may be represented by a Stokes vector of the form

$$S_c^{(0)}(j_0) = \begin{bmatrix} 1 \\ 0 \\ \vdots \\ \vdots \\ 0 \end{bmatrix}_c, \tag{6.36}$$

where we have assumed unit population and the subscript c denotes the reference axes. The intermediate state (j_1, γ_1) is described by the Stokes vector

$$S_c^{(1)}(j_1, \gamma_1) = \bar{Z}_{cc}^{(1)}(j_1, \gamma_1; j_0)S_c^{(0)}(j_0) \tag{6.37}$$

and has elements (taking $\phi_1 = 0$ for simplicity)

$$\{I^{(1)}t_{K'Q'k'q'}(j_1, \gamma_1)\}_c = I_0^{(1)}Z_{00}^{K'Q'k'q'}(j_1, \gamma_1; j_0). \tag{6.38}$$

Second decay. The final state (j_2, γ_2) is described by the Stokes vector

$$S_c^{(2)}(j_2, \gamma_2) = \bar{Z}_{cc}^{(2)}(j_2, \gamma_2; j_1)S_c^{(1)}(j_1),$$

which has elements

$$\{I^{(2)}t_{K''Q''k''q''}(j_2, \gamma_2)\}_c$$
$$= I_0^{(2)} \sum_{K'Q'} Z_{K'Q'}^{K''Q''k''q''}(j_2, \gamma_2; j_1)\{I^{(1)}t_{K'Q'}(j_1)\}_c \, \mathrm{e}^{-\mathrm{i}(Q'-Q''-q'')\phi_2}. \tag{6.39}$$

Thus

$$\{I^{(2)}\}_c = I_0^{(2)} \sum_{K'Q'} Z_{K'Q'}^{0000}(j_2, \gamma_2; j_1)\{I^{(1)}t_{K'Q'}(j_1)\}_c e^{-iQ'\phi_2}$$

$$= I_0^{(2)} \sum_{K'Q'} Z_{K'Q'}^{0000}(j_2, \gamma_2; j_1)\{I^{(1)}t_{K'Q'00}(j_1, \gamma_1)\}_c e^{-iQ'\phi_2}$$

$$= I_0^{(2)}I_0^{(1)} \sum_{K'Q'} Z_{K'Q'}^{0000}(j_2, \gamma_2; j_1)Z_{00}^{K'Q'00}(j_1, \gamma_1; j_0) e^{-iQ'\phi_2}, \quad (6.40)$$

which gives the *directional correlation* of the two gamma rays for axes c. Similarly, the *polarization correlation* in which the polarization of the second radiation is determined is given by

$$\{I^{(2)}t_{k''q''}(\gamma_2)\}_c = I_0^{(2)}I_0^{(1)} \sum_{K'Q'} Z_{K'Q'}^{00k''q''}(j_2, \gamma_2; j_1)Z_{00}^{K'Q'00}(j_1, \gamma_1; j_0) e^{-i(Q'-q'')\phi_2}.$$

$$(6.41)$$

By employing directional and polarization correlation experiments, information may be obtained about the coefficients $Z_{00}^{K'Q'00}(j_1, \gamma_1 : j_0)$ and $Z_{K'Q'}^{00\,k''q''}(j_2, \gamma_2; j_1)$, and hence the nature (spin, parity, etc.) of the nuclear states involved in the transitions.

6.5.2. Particle–gamma correlation

As a second example we consider particle–gamma correlation experiments, i.e. a reaction (R) in which one of the final particles undergoes radiative decay (D), e.g.

$$(R) \ s+t \to s'+t', \qquad (D) \ t' \to t''+\gamma. \quad (6.42)$$

In this case the observation of particle s' along some fixed direction results in the gamma ray showing an angular correlation with respect to this direction. From eqn (5.62) the Stokes vector describing the final reaction products has elements (for unit intensity incident beam)

$$\{I^{(R)}t_{K'Q'k'q'}(s', t')\}_{c'} = X^{(R)} \sum_{KQkq} Z_{KQkq}^{K'Q'k'q'}(s', t'; s, t)\{t_{KQ}(s)t_{kq}(t)\}_c e^{-i(Q+q)\phi_R}.$$

$$(6.43)$$

From eqn (6.33) the Stokes vector for the system (t'', γ) has elements (for axes c')

$$\{I^{(D)}t_{K''Q''k''q''}(t'', \gamma)\}_{c'} = I_0^{(D)} \sum_{K'Q'} Z_{K'Q'}^{K''Q''k''q''}(t'', \gamma; t')\{I^{(R)}t_{K'Q'}(t')\}_{c'}e^{-i(Q''-Q'-q'')\phi_\gamma}.$$

$$(6.44)$$

Consequently,

$$\{I^{(D)}t_{k''q''}(\gamma)\}_{c'} = I_0^{(D)} \sum_{K'Q'} Z_{K'Q'}^{00k''q''}(t'', \gamma; t')\{I^{(R)}t_{00K'Q'}(s', t')\}_{c'} e^{-i(Q'-q'')\phi_\gamma}$$

$$= I_0^{(D)}X^{(R)} \sum_{K'Q'KQkq} Z_{K'Q'}^{00k''q''}(t'', \gamma; t')Z_{KQkq}^{00K'Q'}(s', t'; s, t) \times$$

$$\times \{t_{KQ}(s)t_{kq}(t)\}_c e^{-i(Q'-q'')\phi_\gamma} e^{-i(Q+q)\phi_R}. \quad (6.45)$$

Thus for an unpolarized incident system the *directional correlation* is given by

$$\{I^{(D)}\}_{c'} = I_0^{(D)} X^{(R)} \sum_{K'Q} Z_{K'Q'}^{0000}(t'', \gamma; t') Z_{0000}^{00K'Q'}(s', t'; s, t) e^{-iQ'\phi_\gamma}. \qquad (6.46)$$

The method outlined above for gamma–gamma and particle–gamma correlations is readily seen to be applicable to triple (and more) gamma cascades and to particle–gamma–gamma correlations. For further details the reader should consult a standard text on angular correlations (e.g. Ferguson (1965)) in which the sums over the spin projections, which occur implicitly in the Mueller coefficients, are carried out using Racah algebraic methods as in Section 5.6 for the reaction $t(s, s')t'$.

6.6. Optical pumping

We now discuss the principle of 'optical pumping', i.e. the absorption of polarized light to produce a system of levels (e.g. atomic energy levels in a magnetic field) whose populations are different from the Boltzmann distribution. As a simple example we consider the inverse of the Zeeman effect for a fictitious 'inverted' mercury (even isotope) 2537 Å line, i.e. we assume that the 6^1S_0 level is higher than the 6^3P_1 level (see Fig. 6.2).

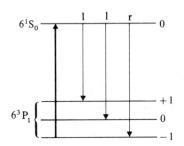

FIG. 6.2. Optical pumping experiment for fictitious 'inverted' $6^1S_0 \rightarrow 6^3P_1$ transition in mercury (even isotope). Absorption of polarized light leads to non-Boltzmann distribution for populations of 6^3P_1 sub-levels following line (l) and resonance (r) fluorescence.

If the appropriate polarized light (the nature of which will be discussed shortly) is incident upon such a 'mercury' vapour cell (under the influence of a magnetic field so that the levels are non-degenerate) only atoms in the sub-level $6^3P_1(-1)$ with magnetic quantum number -1 are excited to the state $6^1S_0(0)$ with magnetic quantum number 0 by the absorption of a photon. The excited state decays at approximately equal rates to each sub-level of the 6^3P_1 state with the emission of *line fluorescence* to the $6^3P_1(+1)$ and $6^3P_1(0)$ sub-levels, and *resonance fluorescence* to the lowest sub-level $6^3P_1(-1)$. Thus the result after some time is a transfer of atoms from the sub-level $6^3P_1(-1)$ to the sub-levels $6^3P_1(+1)$ and $6^3P_1(0)$ since the latter states can only decay relatively slowly to the ground state by the emission of $\mathcal{M}1$ or $\mathcal{E}2$ radiation. The 6^3P_1 sub-levels, which originally were nearly equally populated, may have very different populations after the pumping, the actual result depending upon several factors such as the length of pumping time, intensity of the pumping radiation, etc.

The nature of the required polarized radiation will depend upon the experimental arrangement. For simplicity we only discuss the case in which the light is incident along a direction opposite to that of the magnetic field direction. In this case the *absorption matrix* $\bar{\Delta}_{cc}^A$ corresponding to the inverse process of decay is simply related (assuming reciprocity) to the decay matrix: we have (analogous to eqn (4.85)) for $\theta = 0$ in eqn (6.9)

$$(\bar{\Delta}_{cc}^A)_{m_i\,m_f\mu} \propto (-1)^{m_i - m_f - \mu}(\bar{\Delta}_{cc})_{m_f\mu\,m_i}, \tag{6.47}$$

i.e.

$$\bar{\Delta}_{cc}^A \propto (0\,0\,1\,0\,0\,0\,1\,0\,0)_{cc}. \tag{6.48}$$

Strictly speaking, the absorption matrix is related directly to the *induced emission* part of the decay matrix but since the remaining *spontaneous emission* part of the decay matrix is proportional to the induced emission part (see e.g. Davydov (1965)), relation (6.47) is valid. The form of $\bar{\Delta}_{cc}^A$ implies that only the sub-level $6^3P_1(-1)$ will be excited by absorption of radiation of the form

$$\chi_c^{(\gamma)} = \begin{bmatrix} 1 \\ 0 \\ 0 \end{bmatrix}_c, \tag{6.49}$$

which for the experimental arrangement under discussion corresponds to left-handed circularly polarized light travelling along the negative z-axis, i.e. opposite to the field direction.

6.6.1. Double-resonance method

If 'real' mercury (even isotope) vapour is illuminated with 2537 Å resonance radiation as in the previous section, the appropriate absorption matrix for the process is (setting $\theta = 0$ in relation (6.11) and using relation (6.47))

$$\bar{\Delta}_{cc}^A \propto \begin{bmatrix} 1 & 0 & 0 \\ \hline 0 & 0 & 0 \\ \hline 0 & 0 & 1 \end{bmatrix}_{cc}, \tag{6.50}$$

where the broken lines symbolize the separation of the final 6^3P_1 sub-levels by the magnetic field. The initial state (6^1S_0 atomic level and left-handed circularly polarized light travelling in the opposite direction to the field) may be represented by the direct-product wave function

$$\chi_c^{(i)} \propto (1)_c \otimes \begin{bmatrix} 1 \\ 0 \\ 0 \end{bmatrix}_c = \begin{bmatrix} 1 \\ 0 \\ 0 \end{bmatrix}_c. \tag{6.51}$$

The resultant state is given by

$$\bar{\Delta}_{cc}^A \chi_c^{(i)} \propto 1 \begin{bmatrix} 1 \\ 0 \\ 0 \end{bmatrix}_c + 0 \begin{bmatrix} 0 \\ 1 \\ 0 \end{bmatrix}_c + 0 \begin{bmatrix} 0 \\ 0 \\ 1 \end{bmatrix}_c, \tag{6.52}$$

i.e. only transitions to the $6^3P_1(+1)$ sub-level occur. This sub-level decays to the 6^1S_0 state with the emission of σ-radiation when viewed perpendicular to the magnetic field (see Fig. 6.3).

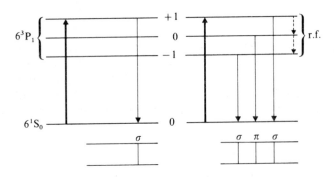

FIG. 6.3. Double-resonance method for mercury (even isotope) $6^3P_1 \rightarrow 6^1S_0$ (2537 Å) transition. Absorption of appropriate polarized light ($M = 1$) in presence of resonant radio-frequency radiation which induces transitions (r.f.) between sub-levels 6^3P_1 leads to additional components in resonance fluorescence.

If an appropriate radio-frequency field is applied to the system, transitions are induced between the magnetic sub-levels of the 6^3P_1 level so that the radiation emitted transversely to the field will contain π- and σ-components resulting from the decay of the $6^3P_1(0)$ and $6^3P_1(-1)$ sub-levels, respectively. The changes in the intensity and polarization of the emitted radiation as a result of magnetic resonance absorption permit the measurement of the radio-frequency resonance and hence the energy differences between the 6^3P_1 sub-levels. For further discussion the reader is referred to the original experiment of Brossel and Bitter (1952).

6.6.2. Energy-level crossing experiment

Another technique which is useful for obtaining information on level separations is the *energy-level crossing experiment*. In this case an external magnetic field of the appropriate strength is applied to the system so that two levels which fluoresce to the same lower level have a separation of the order of their natural line width. As the levels 'cross' the change in the intensity and polarization of the emitted radiation may be detected and the fine- or hyperfine-level splitting may be determined. For a typical experiment see Thaddeus and Novick (1962).

7

RELATIVISTIC TREATMENT
OF POLARIZATION

THE general formalism developed in Chapter 5 for reactions of the type $s+t \rightarrow s'+t'$ is based upon the non-relativistic approximation that the spin of a particle does not depend upon which Lorentz frame of reference is used. In this chapter we indicate the essential changes to the formalism which are required to describe particles moving with relativistic velocities. The case of massless particles is also considered. For convenience in the following sections we adopt the natural units: c (the velocity of light in free space) $= \hbar = 1$.

7.1. Relativistic definition of the spin of a particle with mass

If space–time is homogeneous and isotropic and (for an isolated system) physical processes are independent of the choice of reference frame, the reaction matrix $\hat{M}_{cc}(s', t'; s, t)$ is required to commute with ten operators. These are the *four* operators associated with the infinitesimal displacements of the space–time coordinates: $\hat{P}_\mu \equiv (\hat{\mathbf{P}}, i\hat{H})$, where $\hat{\mathbf{P}} \equiv (\hat{P}_x, \hat{P}_y, \hat{P}_z)$ and $\hat{H} \equiv \hat{P}_t$ may be identified respectively with the total momentum and energy of the system, and the *six* operators corresponding to infinitesimal space–time rotations: $\hat{M}_{\mu\nu} = -\hat{M}_{\nu\mu} \equiv (\hat{\mathbf{J}}, \hat{\mathbf{K}})$, where $\hat{\mathbf{J}} \equiv (\hat{J}_x, \hat{J}_y, \hat{J}_z) \equiv (\hat{M}_{yz}, \hat{M}_{zx}, \hat{M}_{xy})$ and $\hat{\mathbf{K}} \equiv (\hat{K}_x, \hat{K}_y, \hat{K}_z) \equiv (\hat{M}_{xt}, \hat{M}_{yt}, \hat{M}_{zt})$ generate rotations about and Lorentz transformations along the three coordinate axes. For a general discussion of the operators $\hat{P}_\mu$ and $\hat{M}_{\mu\nu}$ the reader is referred to a standard text (e.g. Schweber 1961). The ten operators satisfy the following commutation relations (Foldy 1956):

$$[\hat{P}_i, \hat{P}_j] = 0, \qquad [\hat{P}_i, \hat{H}] = 0, \tag{7.1}$$

$$[\hat{J}_i, \hat{P}_j] = i\varepsilon_{ijk}\hat{P}_k, \qquad [\hat{J}_i, \hat{H}] = 0, \tag{7.2}$$

$$[\hat{J}_i, \hat{J}_j] = i\varepsilon_{ijk}\hat{J}_k, \qquad [\hat{J}_i, \hat{K}_j] = i\varepsilon_{ijk}\hat{K}_k, \tag{7.3}$$

$$[\hat{K}_i, \hat{P}_j] = i\delta_{ij}\hat{H}, \qquad [\hat{K}_i, \hat{H}] = i\hat{P}_i, \qquad [\hat{K}_i, \hat{K}_j] = -i\varepsilon_{ijk}\hat{J}_k, \tag{7.4}$$

where δ_{ij} and ε_{ijk} are the Kronecker and Levi–Civita symbols, respectively:

$$\left. \begin{array}{l} \delta_{ij} = 1 \quad \text{for } i = j \\ \phantom{\delta_{ij}} = 0 \quad \text{otherwise} \end{array} \right\} \quad \left. \begin{array}{l} \varepsilon_{ijk} = 1 \text{ for even permutations of } xyz \\ \phantom{\varepsilon_{ijk}} = -1 \text{ for odd permutations of } xyz \\ \phantom{\varepsilon_{ijk}} = 0 \text{ otherwise} \end{array} \right\}, \tag{7.5}$$

$[\hat{A}, \hat{B}] = \hat{A}\hat{B} - \hat{B}\hat{A}$, and i, j, k denote x, y, or z.

For systems possessing mass, the operator $\hat{\mathbf{J}}$ may be identified with the total angular momentum of the system around its centre of mass $\hat{\mathbf{R}}$ (Foldy 1956; Chou and Shirokov 1958), i.e.

$$\hat{\mathbf{J}} = (\hat{\mathbf{R}} \times \hat{\mathbf{P}}) + \hat{\mathbf{S}}, \tag{7.6}$$

where $\hat{\mathbf{S}}$ is the intrinsic (spin) angular momentum operator and $\hat{\mathbf{R}}$ is given by

$$\hat{\mathbf{R}} = E^{-1}\{\hat{\mathbf{K}} + \tfrac{1}{2}\mathrm{i}\mathbf{P}E^{-1} - (\hat{\mathbf{S}} \times \mathbf{P})(E + M_0)^{-1}\}, \tag{7.7}$$

where we have employed the momentum representation $(\hat{\mathbf{P}} \equiv \mathbf{P})$, $E = (\mathbf{P}^2 + M_0^2)^{\frac{1}{2}}$, and M_0 is the rest mass of the system. The components of $\hat{\mathbf{S}}$ and $\hat{\mathbf{R}}$ satisfy the usual commutation relations for an angular momentum and coordinate operator, respectively:

$$[\hat{S}_i, \hat{S}_j] = \mathrm{i}\varepsilon_{ijk}\hat{S}_k, \qquad [\hat{S}_i, \hat{P}_j] = 0, \qquad [\hat{S}_i, \hat{H}] = 0, \tag{7.8}$$

$$[\hat{R}_i, \hat{R}_j] = 0, \qquad [\hat{R}_i, \hat{P}_j] = \mathrm{i}\delta_{ij}, \qquad [\hat{R}_i, \hat{H}] = \mathrm{i}\hat{P}_i E^{-1} \qquad [\hat{R}_i, \hat{S}_j] = 0. \tag{7.9}$$

For a single particle moving with momentum $\mathbf{P}$, the operators $\hat{\mathbf{R}}$ and $\hat{\mathbf{S}}$ represent the 'position' of the particle and its angular momentum relative to its centre of mass $\hat{\mathbf{R}}$, respectively. Thus $\hat{\mathbf{S}}$ is the operator associated with the intrinsic spin of the particle. The commutation relations $[\hat{S}_i, \hat{H}] = [\hat{J}_i, \hat{H}] = 0$ imply that the orbital and spin angular momenta are separately conserved. Consequently, for spin-$\frac{1}{2}$ particles, the operators $\hat{\mathbf{S}}$ and $\hat{\mathbf{R}}$ do *not* correspond to the spin and coordinate operators usually associated with the Dirac equation but are to be identified more closely with the 'mean-spin' and 'mean-position' operators introduced by Foldy and Wouthuysen (1950).

7.2. Spin of a particle with mass in different Lorentz frames

In order to obtain the relationship between the spin operator $\hat{\mathbf{S}}$ in one frame of reference c and the corresponding spin operator $\hat{\mathbf{S}}'$ in a new system c' which moves relative to c with the velocity $\boldsymbol{\beta}$ (in units of the speed of light), we follow the treatment of Chou and Shirokov (1958) and form the four-vector operator

$$\hat{\Gamma}_\kappa = \frac{1}{2\mathrm{i}}\varepsilon_{\mu\nu\kappa\lambda}\hat{M}_{\mu\nu}\hat{P}_\lambda \equiv (\hat{\boldsymbol{\Gamma}}, \hat{\Gamma}_4), \tag{7.10}$$

where $\varepsilon_{\mu\nu\kappa\lambda}$ (analogous to ε_{ijk}) is the Levi–Civita symbol of the fourth rank. In momentum representation

$$\hat{\boldsymbol{\Gamma}} = M_0\hat{\mathbf{S}} + \frac{(\mathbf{P} \cdot \hat{\mathbf{S}})\mathbf{P}}{(E + M_0)}, \qquad \hat{\Gamma}_4 = \mathrm{i}(\mathbf{P} \cdot \hat{\mathbf{S}}), \tag{7.11}$$

which imply that $\hat{\mathbf{S}}^2$ is an invariant quantity under Lorentz transformations since

$$\hat{\boldsymbol{\Gamma}} \cdot \hat{\boldsymbol{\Gamma}} + \hat{\Gamma}_4^2 = M_0^2\hat{\mathbf{S}}^2 \tag{7.12}$$

is invariant. Under a Lorentz transformation ($\boldsymbol{\beta}$) the four-vector operator $\{\hat{\boldsymbol{\Gamma}}, \hat{\Gamma}_4\}$ is transformed into a new four-vector operator $\{\hat{\boldsymbol{\Gamma}}', \hat{\Gamma}_4'\}$, where (Møller 1952)

$$\hat{\boldsymbol{\Gamma}}' = \hat{\boldsymbol{\Gamma}} + \boldsymbol{\beta}\{(\hat{\boldsymbol{\Gamma}} \cdot \boldsymbol{\beta})(\gamma_\beta - 1)\beta^{-2} + i\gamma_\beta \hat{\Gamma}_4\}, \tag{7.13}$$

$$\hat{\Gamma}_4' = \gamma_\beta\{\hat{\Gamma}_4 + i(\hat{\boldsymbol{\Gamma}} \cdot \boldsymbol{\beta})\}, \tag{7.14}$$

and $\gamma_\beta = (1 - \beta^2)^{-\frac{1}{2}}$. If $\hat{\mathbf{S}}'$, $\mathbf{P}'$, E' denote the spin operator, momentum, and energy in the new reference frame then, analogously to eqn (7.11),

$$\hat{\boldsymbol{\Gamma}}' = M_0 \hat{\mathbf{S}}' + \frac{(\mathbf{P}' \cdot \hat{\mathbf{S}}')\mathbf{P}'}{(E' + M_0)}, \qquad \hat{\Gamma}_4' = i(\mathbf{P}' \cdot \hat{\mathbf{S}}'), \tag{7.15}$$

and hence we obtain a relationship between $\hat{\mathbf{S}}'$ and $\hat{\mathbf{S}}$ of the form

$$M_0 \hat{\mathbf{S}}' + \frac{(\mathbf{P}' \cdot \hat{\mathbf{S}}')\mathbf{P}'}{(E' + M_0)} = M_0 \hat{\mathbf{S}} + \frac{(\mathbf{P} \cdot \hat{\mathbf{S}})\mathbf{P}}{(E + M_0)} + \boldsymbol{\beta}(...), \tag{7.16}$$

where

$$(\mathbf{P}' \cdot \hat{\mathbf{S}}') = \gamma_\beta \left\{ (\mathbf{P} \cdot \hat{\mathbf{S}}) - M_0(\boldsymbol{\beta} \cdot \hat{\mathbf{S}}) - \frac{(\mathbf{P} \cdot \hat{\mathbf{S}})(\boldsymbol{\beta} \cdot \mathbf{P})}{(E + M_0)} \right\}, \tag{7.17}$$

$$\mathbf{P}' = \mathbf{P} + \boldsymbol{\beta}(...), \tag{7.18}$$

and

$$E' = \gamma_\beta\{E - (\boldsymbol{\beta} \cdot \mathbf{P})\}. \tag{7.19}$$

Choosing the convenient set of axes—z parallel to $\boldsymbol{\beta}$ and y parallel to $(\boldsymbol{\beta} \times \mathbf{P})$—we obtain immediately from eqn (7.16) that $\hat{S}_y' = \hat{S}_y$. Since $\hat{S}'^2 = \hat{S}^2$, the operator $\hat{\mathbf{S}}'$ is obtained from the operator $\hat{\mathbf{S}}$ by a rotation about the y-axis, and we may write

$$\left. \begin{array}{l} \hat{S}_x' = \hat{S}_x \cos \Omega + \hat{S}_z \sin \Omega \\ \hat{S}_z' = -\hat{S}_x \sin \Omega + \hat{S}_z \cos \Omega \end{array} \right\}. \tag{7.20}$$

The counter-clockwise rotation Ω of the vector $\hat{\mathbf{S}}$ for stationary axes may be determined by considering the x-component of $\hat{\mathbf{S}}'$ and equating the coefficient of $\hat{S}_z$ to $\sin \Omega$ as required by eqn (7.20). We have, from eqns (7.16) and (7.17),

$$\hat{S}_x' = \hat{S}_x + \frac{(P_x^2 \hat{S}_x + P_x P_z \hat{S}_z)}{M_0(E + M_0)} - \gamma_\beta \frac{(P_x^2 \hat{S}_x + P_x P_z \hat{S}_z)}{M_0(E' + M_0)} +$$
$$+ \frac{\beta \gamma_\beta P_x \hat{S}_z}{(E' + M_0)} + \gamma_\beta \beta \frac{(P_x^2 \hat{S}_x + P_x P_z \hat{S}_z)}{M_0(E + M_0)(E' + M_0)}, \tag{7.21}$$

and hence

$$\sin \Omega = \frac{P_x P_z}{M_0(E+M_0)} - \frac{\gamma_\beta P_x P_z}{M_0(E'+M_0)} +$$

$$+ \frac{\gamma_\beta \beta P_x}{(E'+M_0)} + \frac{\gamma_\beta \beta P_x P_z^2}{M_0(E+M_0)(E'+M_0)}. \tag{7.22}$$

Substituting $P_x = |\mathbf{P}| \sin \theta = \gamma M_0 u \sin \theta$, $P_z = \{E - E'/\gamma_\beta\}/\beta$ (from eqn (7.19)), $E = \gamma M_0$, $E' = \gamma' M_0$ into eqn (7.22) yields, after a little algebra,

$$\sin \Omega = \frac{\gamma_\beta \gamma \beta u \sin \theta (1+\gamma+\gamma'+\gamma_\beta)}{(1+\gamma)(1+\gamma')(1+\gamma_\beta)}, \tag{7.23}$$

which gives the rotation of the spin state arising from relativistic effects.

Eqn (7.23) implies that the only essential change in the non-relativistic formalism is an *additional rotation* Ω for each reaction process; i.e. eqn (5.48) for the standard reaction matrix becomes

$$\tilde{M}_{c'c}(s', t'; s, t) = (X/I_0)^{\frac{1}{2}} \{D^{(s')}(\phi, \theta_{\text{lab}} + \Omega, 0)^\dagger \otimes D^{(t')}(\phi, \theta_{\text{lab}} + \Omega, 0)^\dagger\} \times$$

$$\times \hat{M}_{cc}(s', t'; s, t), \tag{7.24}$$

where $\hat{M}_{cc}(s', t'; s, t)$ is the reaction matrix in the c.m. frame c. The relativistic rotation Ω is a consequence of the transformation from c.m. to lab. axes for the final system. It should be noted that no rotation occurs for lab. to c.m. transformations for the initial pair of particles, since $\mathbf{u}$ is parallel to $\boldsymbol{\beta}$. Similarly, in the change from the c.m. system to the rest frame of a particle, $\mathbf{P}$ is parallel to $\boldsymbol{\beta}$ and $\Omega = 0$; it is more convenient to consider the polarization state of a particle as being defined in its rest frame rather than the c.m. frame, which depends upon the nature of the other particle.

7.3. Helicity representation

We now introduce an alternative procedure for describing the polarization of particles. This method proposed by Jacob and Wick (1959) employs the projections of the spins along their respective directions of motion rather than the spin components along some arbitrary axis and is called the *helicity representation*. This treatment has the advantage that it also applies to massless particles.

7.3.1. Particles with mass

The helicity operator $\hat{\Lambda}$ corresponding to particles of spin-s and momentum $\mathbf{k}$ is defined by

$$\hat{\Lambda} = (\hat{\mathbf{S}} \cdot \mathbf{k})/|\mathbf{k}|, \tag{7.25}$$

where $\hat{\mathbf{S}}$ is the intrinsic angular momentum operator of Section 7.1. Thus particles with mass can exist in $(2s+1)$ basic helicity states with eigenvalues of the operator $\hat{\Lambda}$, $\lambda = -s, -s+1, ..., s$. Since $\hat{\Lambda}$ is a scalar operator, it has the convenient property of being invariant under ordinary rotations, and hence the helicity of a particle does not change under a rotation of axes. However, it should be noted that helicity (a) is undefined for a particle at rest and (b) changes sign for a Lorentz transformation which reverses the direction of the momentum $\mathbf{k}$.

For the special case in which the z-axis is along the direction of particle motion, $\hat{\Lambda} \equiv \hat{S}_z$ and the basic helicity states $\phi_c^{(\lambda_s)}$ may be chosen to have the same relative phases as the corresponding spin wave functions with projections m_s, i.e.

$$\phi_c^{(\lambda_s)} \equiv \phi_c^{(m_s)} \quad \text{for } \mathbf{k} = k\mathbf{e}_z. \tag{7.26}$$

Since the standard scattering matrices for spin-$\frac{1}{2}$ and spin-1 particles described in Chapters 3 and 4 are defined with respect to coordinates c, c' in which the z-, z'-axes are along the directions of the incident and final beams of particles, respectively, these scattering matrices may equally well be considered to be in the helicity representation. It is for this reason that the standard axes have been termed helicity coordinate frames. Thus, although for arbitrary axes the helicity formalism leads to simpler expressions for the scattering amplitudes (Jacob and Wick 1959); for the standard axes and the case of spinless targets the expressions are essentially the same.

For the more general reaction $t(s, s')t'$ involving targets with spin we can define two-particle helicity states for the initial and final channels, provided we employ c.m. reference systems. These are of the direct-product form

$$\phi_c^{(\lambda_s)} \otimes \phi_c^{(-\lambda_t)} \equiv \phi_c^{(m_s)} \otimes \phi_c^{(m_t)}, \tag{7.27}$$

and consequently the corresponding reaction matrix has elements

$$M_{\lambda_{s'} - \lambda_{t'} \lambda_s - \lambda_t} \equiv M_{m_{s'} m_{t'} m_s m_t} \tag{7.28}$$

For the case of three (or more) particle final states there is no simple relationship between the helicity and spin projection representations unless one employs *individual* particle helicity frames, i.e. a different coordinate system for each outgoing particle. Thus for the reaction

$$s + t \rightarrow s' + t' + u', \tag{7.29}$$

the standard reaction matrix corresponding to eqn (7.24), with individual helicity coordinate systems for all final particles c_i' ($\equiv c_{s'}' c_{t'}' c_{u'}'$), is given by

$$\tilde{M}_{c_i c}(s', t', u'; s, t) = (X/I_0)^{\frac{1}{2}} \{ D^{(s')}(\phi^{(s')}, \theta_{lab}^{(s')} + \Omega^{(s')}, 0)^\dagger \otimes$$

$$\otimes D^{(t')}(\phi^{(t')}, \theta_{lab}^{(t')} + \Omega^{(t')}, 0)^\dagger \otimes$$

$$\otimes D^{(u')}(\phi^{(u')}, \theta_{lab}^{(u')} + \Omega^{(u')}, 0)^\dagger \} \hat{M}_{cc}(s', t', u'; s, t), \tag{7.30}$$

where $(X/I_0) = d\Omega_{cm}^{(s')}/d\Omega_{lab}^{(s')}$ is the ratio of c.m. and lab. solid angles for the final beam of s' particles; $(\theta_{lab}^{(q')}, \phi^{(q')})$ are the Euler angles required to rotate the initial c.m. axes c, so that the final lab. axes $c_{q'}'$ have the z'-axis along the outgoing beam (q') and the y'-axis parallel to $\mathbf{k} \times \mathbf{k}'(q')$; and $\Omega^{(q')}$ are the corresponding relativistic rotations given by eqn (7.23). For these axes (c_i', c) we have in terms of helicities the standard matrix elements:

$$\tilde{M}_{\lambda_{s'} \lambda_{t'} \lambda_{u'} \lambda_s - \lambda_t} \equiv \tilde{M}_{m_{s'} m_{t'} m_{u'} m_s m_t}, \qquad (7.31)$$

where it is assumed that λ_t remains invariant in the c.m. to lab. Lorentz transformation for the initial axes.

7.3.2. Massless particles

In the limit of M_0 tending to zero, eqn (7.17) implies that

$$\hat{\Lambda}' = \frac{(\mathbf{P}' \cdot \mathbf{S}')}{P'} \rightarrow \gamma_\beta \frac{(\mathbf{P} \cdot \mathbf{S})}{|\mathbf{P}|} \frac{|\mathbf{P}|}{|\mathbf{P}'|} \left\{ 1 - \frac{(\boldsymbol{\beta} \cdot \mathbf{P})}{E} \right\} = \hat{\Lambda}, \qquad (7.32)$$

since $|\mathbf{P}| \rightarrow E$ and $|\mathbf{P}'| \rightarrow E' = \gamma_\beta(E - \boldsymbol{\beta} \cdot \mathbf{P})$. Thus for massless particles the helicity operator (and hence the helicity) is invariant under an *arbitrary* 'proper' Lorentz transformation of the form given by eqns (7.13) and (7.14). This contrasts with the case of particles with mass, for which helicity is conserved only for proper Lorentz transformations which are along the direction of motion and do not change this direction.

If the particles also possess a definite parity $\eta(\pm 1)$ then for the 'improper' Lorentz transformations corresponding to the space-reflection symmetry operator $\hat{\Pi}$, the basic helicity state $\phi_c^{(\lambda)}$ is transformed into the state of opposite helicity

$$\hat{\Pi} \phi_c^{(\lambda)} \propto \eta \phi_c^{(-\lambda)}. \qquad (7.33)$$

Thus for massless spin-s particles with definite parity, the pairs of helicity states $\lambda = \pm s, \pm(s-1), \ldots$ form irreducible sub-groups for the complete set of proper and improper Lorentz transformations. Indeed the photon may be identified with a particle of parity $\eta = -1$, spin $s = 1$, and helicity $\lambda = \pm 1$. These properties require that the photon rest mass is exactly zero and that the associated field equations (Maxwell's) satisfy certain 'gauge invariances' (Corson 1953). If a particle has no definite parity (e.g. a neutrino) then it need only exist in one helicity state. This complete polarization of a particle may lead to remarkable polarization effects in its interaction with other particles. In the following section we discuss two such phenomena observed in weak interaction decay processes involving antineutrinos. For further discussion of the above concepts the reader is referred to the texts by Roman (1960) and Schwinger (1970).

7.4. Some weak interaction processes

In this section we describe two weak interaction decay processes:
(1) the initial parity non-conservation experiment of Wu *et al.* (1957);
(2) the branching ratio $(\pi^- \to e^- + \bar{v}_e)/(\pi^- \to \mu^- + \bar{v}_\mu)$ for negative pion decay;

as a consequence of the assumption that the emitted antineutrinos are completely polarized along their directions of motion, i.e. have helicity $\lambda = +\frac{1}{2}$.

7.4.1. *Beta decay of a polarized nucleus*

Previously, we have restricted our discussion to interactions which conserve parity, e.g. the pure transverse polarization of a beam of spin-$\frac{1}{2}$ particles produced by elastic scattering of an unpolarized incident beam from a spinless target, which is a consequence of parity conservation for the strong nuclear and electromagnetic interactions as described in Chapter 3. We now consider the beta decay of nuclei in which the weak interactions do not conserve parity and the angular distribution of the emitted electrons has the form (Lee and Yang 1956)

$$I(\theta) \propto (1 + \alpha \mathbf{p} \cdot \mathbf{v}) = (1 + \alpha |\mathbf{p}| |\mathbf{v}| \cos \theta), \tag{7.34}$$

where $\mathbf{p} = \langle \hat{\mathbf{j}} \rangle / j$ is the vector polarization of the parent spin-j nuclei, $\mathbf{v}$ is the electron velocity, and $\alpha \neq 0$. Experiments indicate that the coefficient $\alpha \simeq -1$, i.e. that the electrons are emitted preferentially in a direction opposite to that of the magnetic field $\mathbf{H} \equiv H\mathbf{e}_z$ which defines the initial polarization reference direction z. A value of $\alpha = -1$ is consistent with (1) a non-derivative local four-fermion interaction of the (V–A) form with equal amounts of parity conserving and parity violating terms, and (2) antineutrinos existing in only the one helicity state $\lambda = +\frac{1}{2}$ (for details see Konopinski (1959) and Marshak, Riazuddin, and Ryan (1970)). Alternatively, the assumption of a completely polarized antineutrino leads (by consideration of the various experimental data) naturally to the (V–A) type of current–current interaction, which implies $\alpha = -1$ in eqn (7.34).

In the experiment of Wu *et al.* (1957), the beta decay of oriented ^{60}Co nuclei:

$$^{60}\vec{\text{Co}} \to {}^{60}\text{Ni} + \vec{e}^- + \bar{v}_e \tag{7.35}$$
$$(j_i = 5) \quad (j_f = 4)$$

was studied, and the asymmetry of the electron distribution

$$\varepsilon = \{I(\theta = 0) - I(\theta = \pi)\}/\{I(\theta = 0) + I(\theta = \pi)\} = \alpha |\mathbf{p}| |\mathbf{v}| \tag{7.36}$$

was found to be large and negative. In particular, the coefficient α was determined to be about -1. In terms of the complete polarization of the antineutrino, this result for α can be understood as follows.

For simplicity let us consider a completely vector polarized ^{60}Co system with $|\mathbf{p}| = p_z = 1$. In this case the decay (7.35) is essentially a pure allowed (i.e. zero orbital angular momentum is carried away by the leptonic pair) Gamow–Teller transition ($m_i = 5 \rightarrow m_f = 4$) involving one unit ($m = 1$) of the component of the spin angular momentum along the field direction $\mathbf{H}$ for the electron–antineutrino system. This means that the individual spin components of the electron and antineutrino are required to be parallel to one another in order to conserve angular momentum (see Fig. 7.1). If the anti-neutrino has helicity $\lambda = +\frac{1}{2}$ it is required to travel along the field direction and consequently the associated electron which carries away most of the residual linear momentum tends to be emitted at $\theta = \pi$.

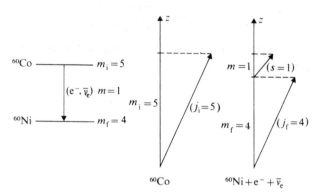

FIG. 7.1. Beta decay of completely vector-polarized ^{60}Co nuclei with $|\mathbf{p}| = p_z = 1$ as a pure allowed Gamow–Teller transition ($m_i = 5 \rightarrow m_f = 4$). Conservation of angular momentum requires parallel spin components for electron and antineutrino.

7.4.2. *Negative pion decay*

The two-body leptonic decays of the negative pion are

$$(1) \ \pi^- \rightarrow \mu^- + \bar{\nu}_\mu \quad \text{and} \quad (2) \ \pi^- \rightarrow e^- + \bar{\nu}_e. \tag{7.37}$$

Although process (2) is favoured energetically, the decay mode (1) is dominant. This can be understood in terms of the complete polarization ($\lambda_{\bar{\nu}} = +\frac{1}{2}$) of both kinds of antineutrinos as follows.

In the c.m. system the charged lepton and antineutrino have exactly opposite linear momenta. Thus, since the pion is spinless and the weak interaction decay process is essentially a pure allowed Fermi transition ($j_i = 0 \rightarrow j_f = 0$), conservation of angular momentum requires the charged lepton to have the same helicity $\lambda_l = +\frac{1}{2}$ as the associated antineutrino (see Fig. 7.2). Now for $v \simeq 1$ the (V–A) interaction strongly favours a decay into *opposite* helicity states for the leptonic pair (Konopinski 1959; Marshak *et al.* 1970) and consequently the two-body leptonic decays (1) and (2) of the negative pion are both inhibited. However, because the mass m_e of the

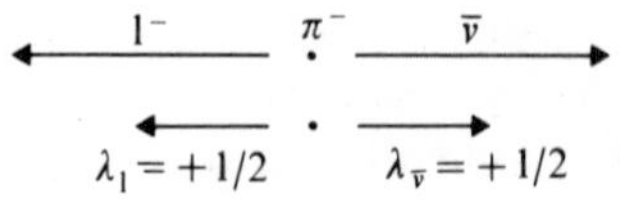

FIG. 7.2. Two-body leptonic decay of negative pion ($\pi^- \to l^- + \bar{\nu}$) in c.m. system as a pure allowed Fermi transition ($j_i = 0 \to j_f = 0$). Conservation of angular momentum requires anti-parallel spin components for leptonic pair.

electron is much smaller than the muonic mass m_μ, the decay mode (2) tends to be more strongly inhibited than the mode (1). Indeed, assuming equality of the pion decay coupling constants for both processes, the branching ratio R is given by (Marshak *et al.* 1970)

$$R = \frac{(\pi^- \to e^- + \bar{\nu}_e)}{(\pi^- \to \mu^- + \bar{\nu}_\mu)} = \frac{m_e^2(m_\pi^2 - m_e^2)}{m_\mu^2(m_\pi^2 - m_\mu^2)} \simeq 10^{-4}, \tag{7.38}$$

where m_π is the negative pion mass. This value of R is in good agreement with experiment.

REFERENCES

BARSCHALL, H. H. and HAEBERLI, W. (eds) (1971). *Proceedings of the third international symposium on polarization phenomena in nuclear reactions.* University of Wisconsin Press, Madison.

BEROVIC, N. (1970). *Nucl. Phys.* **A157**, 106.

BIEDENHARN, L. C. (1959). *Nucl. Phys.* **10**, 620.

BOHR, A. (1959). *Nucl. Phys.* **10**, 486.

BRINK, D. M. and SATCHLER, G. R. (1968). *Angular momentum* (2nd edn). Clarendon Press, Oxford.

BROSSEL, J. and BITTER, F. (1952). *Phys. Rev.* **86**, 308.

CHOU, K. and SHIROKOV, M. I. (1958). *Soviet Phys. J.E.T.P.* **7**, 851. (Trans. from *Zh. éksp. teor. Fiz.* **34**, 1230.)

COESTER, F. (1951). *Phys. Rev.* **84**, 1259.

CORSON, E. M. (1953). *Introduction to tensors, spinors and relativistic wave equations.* Blackie and Sons, Glasgow.

CONDON, E. U. and SHORTLEY, G. H. (1935). *The theory of atomic spectra.* Cambridge University Press, Cambridge.

CSONKA, P. L., MORAVSCIK, M. J., and SCADRON, M. D. (1966). *Ann. Phys.* **40**, 100.

DAVYDOV, A. S. (1965). *Quantum mechanics.* Pergamon Press, Oxford.

EMMERSON, J. Mc. (1972). *Symmetry principles in particle physics.* Clarendon Press, Oxford.

FANO, U. (1949). *J. opt. Soc. Am.* **39**, 859.

——— (1957). *Rev. mod. Phys.* **29**, 74.

FERGUSON, A. J. (1965). *Angular correlation methods in gamma-ray spectroscopy.* North-Holland, Amsterdam.

FOLDY, L. L. (1956). *Phys. Rev.* **102**, 568.

——— and WOUTHUYSEN, S. A. (1950). *Phys. Rev.* **78**, 29.

GOLDFARB, L. J. B. (1958). *Nucl. Phys.* **7**, 622.

HODGSON, P. E. (1963). *Optical model of elastic scattering.* Clarendon Press, Oxford.

——— (1966). *Adv. Phys.* **15**, 329.

JACOB, M. and WICK, G. C. (1959). *Ann. Phys.* **7**, 404.

JONES, R. C. (1941). *J. opt. Soc. Am.* **31**, 488.

KONOPINSKI, E. J. (1959). *A. Rev. nucl. Sci.* **9**, 99.

LAKIN, W. (1955). *Phys. Rev.* **98**, 139.

LEE, T. D. and YANG, C. N. (1956). *Phys. Rev.* **104**, 254.

MARSHAK, R. E., RIAZUDDIN, and RYAN, C. P. (1969). *Theory of weak interactions in particle physics.* Wiley, New York.

MØLLER, C. (1952). *The theory of relativity.* Clarendon Press, Oxford.

OHLSEN, G. G. (1972). *Rep. Prog. Phys.* **35**, 717.

POINCARÉ, H. (1892). *Théorie mathematique de la lumière* Vol. 2. Corré, Paris.

ROMAN, P. (1960). *Theory of elementary particles.* North-Holland, Amsterdam.

ROSE, M. E. (1955). *Multipole fields.* Wiley, New York.

——— (1957). *Elementary theory of angular momentum.* Wiley, New York.

SATCHLER, G. R. (1960). *Nucl. Phys.* **21**, 116.

SCHWEBER, S. S. (1961). *An introduction to relativistic quantum field theory*. Row, Peterson, Evanston, Illinois.

SCHWINGER, J. (1970). *Particles, sources and fields*. Addison-Wesley, Reading, Massachusetts.

SHORE, B. W. and MENZEL, D. H. (1968). *Principles of atomic spectra*. Wiley, New York.

SHURCLIFF, W. A. (1962). *Polarized light: production and use*. Harvard University Press, Cambridge, Massachusetts.

STOKES, G. G. (1852). *Trans. Camb. phil. Soc., Math. phys. Sci.* **9**, 399.

THADDEUS, P. and NOVICK, R. (1962). *Phys. Rev.* **126**, 1774.

TOLHOEK, H. A. and GROOT, de S. R. *Physica* **17**, 1.

WAERDEN, van der B. L. (1960). *Theoretical physics in the twentieth century* (ed. M. Fierz and V. F. Weisskopf), Interscience, New York.

WOLFENSTEIN, L. (1954). *Phys. Rev.* **96**, 1654.

WU, C. S., AMBLER, E., HAYWARD, R. W., HOPPES, D. D., and HUDSON, R. P. (1957). *Phys. Rev.* **105**, 1413.

AUTHOR INDEX

SUBJECT INDEX